D0945257

# STATISTICS
## Tool
## of the Behavioral Sciences

# STATISTICS
## Tool
## of the Behavioral Sciences

MARCIA K. JOHNSON
ROBERT M. LIEBERT

*State University of New York*
*at Stony Brook*

PRENTICE-HALL, INC., ENGLEWOOD CLIFFS, NEW JERSEY 07632

*Library of Congress Cataloging in Publication Data*

JOHNSON, MARCIA K.
Statistics.

Includes index.
1. Statistics. I. Liebert, Robert M., joint author. II. Title.
HA29.J57 519.5 76-43557
ISBN 0-13-844704-7



Printed in the United States of America

10 9 8 7 6 5 4 3 2 1

*Prentice-Hall International, Inc., London*
*Prentice-Hall of Australia Pty. Limited, Sydney*
*Prentice-Hall of Canada, Ltd., Toronto*
*Prentice-Hall of India Private Limited, New Delhi*
*Prentice-Hall of Japan, Inc., Tokyo*
*Prentice-Hall of Southeast Asia Pte. Ltd., Singapore*
*Whitehall Books Limited, Wellington, New Zealand*

*To Geoffrey Keppel and Keith Clayton,*
*who contributed greatly*
*to our own understanding of statistics*

# Contents

## 6

## DESIGNING POWERFUL EXPERIMENTS 75



## 7

## REPEATED-MEASURES AND MATCHED-PAIRS DESIGNS 91

## 8

## VARIANCE RATIOS: THE F TEST 104



## 9

## COMPARISONS AMONG TREATMENT MEANS IN SINGLE-FACTOR EXPERIMENTS 119

## 10

## FACTORIAL DESIGNS: BASIC CONCEPTS 135



## 11

## REGRESSION AND CORRELATION 159



## 12

## NONPARAMETRIC STATISTICS: DEALING WITH FREQUENCIES AND RANKS 179

# Preface

Few courses arouse more apprehension among beginning students than statistics. The reasons for this are easy to understand. Statistics seems to be an abstract topic with little for those who are not mathematically oriented to hold on to and often seems to be a needless requirement to impose on a social science or education major. Unfortunately, many students finish a statistics course with essentially these same feelings.

Actually, few courses are more relevant to developing skills in critical thinking and a general understanding of how the world works. Statistical procedures are very powerful tools which, properly applied, allow professional scientists and non-scientists alike to answer clearly some of the most important questions about truth, accuracy, and certainty which arise in trying to understand the world. Statistical tests are based on a set of well-defined and reasonable rules for decision making that can be understood and applied by almost anyone. The stumbling block for many introductory students is that statisticians are often most comfortable illustrating the elegance of mathematical proofs and describing the procedures they use in the special language of mathematical notation, both of which present difficulties for students without a strong mathematical background. Interest in a potentially useful topic is therefore lost quickly or never won in the first place.

In this book we have tried to overcome some of these problems by introducing statistical concepts in the context of the overall process of scientific inquiry. In addition, the reader is shown that statistics are or can be used to answer questions and resolve practical issues that arise in everyday life. We begin by introducing a small number of methodological concepts central to how questions are asked and answered by scientists and build the fundamental ideas of statistics—samples, sampling distributions, and the concept of statistical inference—around specific research problems that have a logic that can be followed easily. Mathematical proofs are de-emphasized, and key concepts are examined both verbally and graphically. Likewise, formal notation is kept to a minimum and always introduced with complete verbal translations. Practice problems and examples are interspersed throughout the book to facilitate understanding the application of statistics to interesting hypothetical and actual problems.

Many students at the State University of New York at Stony Brook have helped in the development of this book. We are especially grateful to the students in our course in the Fall, 1975, who used a draft of the book as their only text. We feel the manuscript benefited greatly from this classroom testing. We would also like to acknowledge the many helpful comments of Harvey Blumberg, Morton Friedman, Paul Isaac, Robert Porter, and Robert Young.

MARCIA K. JOHNSON
ROBERT M. LIEBERT
*Stony Brook, N.Y.*

# STATISTICS
## Tool
## of the Behavioral Sciences

# Statistics as a Tool

# 1

LIKE MANY OTHER HUMAN ACTIVITIES, the goal of science is a better understanding of the world we live in. For scientists, this understanding is not simply personal; it must be shared. To become a part of scientific knowledge, one scientist's discovery must be accepted by other scientists. And to be accepted, knowledge must be acquired according to certain rules or conventions. This book is concerned primarily with these conventions and rules. As ideas, the rules themselves are not hard to understand, for they are based on the kind of disciplined common sense that intelligent people try to apply to situations confronting them every day.

For example, here are some statements similar to those in advertisements that have appeared recently in magazines or newspapers. As you read each, think about the assumptions that the advertiser wants you to make, the conclusions he wants you to reach, and the information that you, as a critical reader, would want to know before you would believe what you are expected to believe.

1. "A car just drained of our new Pistonclean oil went a distance of over 60 miles protected by just the thin film of lubrication left in the engine. With absolutely no engine damage!"

2. "80 percent of all major tournament winners last year used Victory golf balls."
3. "9 out of 10 Medallion automobiles sold in the United States in the last ten years are still on the road!"
4. "Breathless cigarettes are lower in both tar and nicotine than 99 percent of all other cigarettes sold."

The statements are presumably true but are not complete by themselves. In each case, the statement also requires an *inference* by you, the potential consumer. Before drawing the inference, then, you must be certain that it is valid; this in turn raises additional questions. For example:

1. *Inference:* Most cars, including my car, would go 60 miles with just a thin film of Pistonclean; therefore, Pistonclean is a good oil to buy.
   *Additional questions:* But did they test just one car? What kind of car did they use? Would any other brand of oil do the same or more? Could a car that had never had oil in it at all also go 60 miles with no damage?
2. *Inference:* If most winners use them, Victory balls must be good for your golf game.
   *Additional questions:* How many *losers* use Victory golf balls? More than 80 percent? Would the winners who use Victory play just as well using other brands? (Maybe Victory gives them the biggest discount.)
3. *Inference:* Most Medallions (90 percent) last for at least ten years. Good car.
   *Additional questions:* How many Chevys, Fords, and Plymouths sold in the last ten years are still on the road? Were 90 percent of the Medallions sold in the United States in the last ten years sold only last year? Even if the Medallions do not tend to be newer than other cars, were they purchased by the same types of people? Were they driven under the same road conditions for an equivalent number of miles? Were they repaired or serviced more often?
4. *Inference:* Breathless is about the lowest cigarette in tar and nicotine you can buy.
   *Additional questions:* How many cigarettes of each brand did they test? Is there very much variation in the amount of tar and nicotine from cigarette to cigarette within any particular brand? Even if Breathless cigarettes do contain less tar and nicotine, how much less? Do they still contain enough to damage my health?

All these examples of advertisements use statistics. That is, they include numbers that are descriptions of single observations (the car went over 60 miles) and numbers that summarize many observations (80 percent of all winners use Victory). One of the major purposes in collecting and summarizing these observations was to suggest inferences about similar items or events that were not observed or measured (any car would go farther with no damage when protected by a thin film of Pistonclean) and to suggest implications (you will live longer if you smoke Breathless) based on these "facts."

When is it safe to make inferences about a general state of affairs after observing a few instances? And what are reasonable implications of this general state of affairs? You might feel that one has to be particularly careful in reading advertisements because, after all, it is an advertiser's job to get you to believe what he wants you to regardless of whether or not it is "true." However, you might also maintain that in most other cases the facts speak for themselves. As we will see, this is rarely the case.

The problem of what to believe is not confined to situations in which you may have some reason for thinking that someone might want to put something over on you. Much of the information we receive daily (even in some advertisements) is offered in good faith or is based on our own observations. Even in these cases, it is not always clear how to tell the difference between reasonable and unreasonable inferences. Suppose you read in the science section of your newspaper that a new treatment for cancer produced a 60 percent cure rate, whereas the cure rate for the old treatment is usually reported at 40 percent. Is this information sufficient grounds for concluding that the new treatment is superior to the old treatment? Not necessarily. People vary in their responsiveness to medication and in the particular way in which their bodies function. The treatment's apparent success might reflect a fortuitous or "chance" result that does not guarantee its general superiority in treating the rest of the population. How likely are the patients treated with the new medicine to be characteristic of cancer patients in general?

This example is not a unique case. In the social and behavioral sciences, most of what we know and believe is based on inferences about large groups (e.g., all schoolchildren, all prejudiced people, all women) drawn from information and observations of much smaller groups. To avoid erroneous conclusions, we turn to statistics as a tool to help evaluate the facts before us.

Of course, *statistics* is a very broad term. The researcher actually turns to specific statistical tests (coupled with certain important principles of research design) to provide guidelines both for drawing conclusions from limited information and for gathering information in the first place. Our purpose in the pages that follow is to introduce the basic principles of statistical analysis and research design in a fashion that both emphasizes an understanding of how they work and provides clear guidelines for their use and application. It will be helpful to begin with a preview of the research enterprise that statistical tools are designed to serve.

## STATISTICS AND THE RESEARCH ENTERPRISE

It is impossible to offer a comprehensive definition of research. However, the idea of research usually carries with it the notion that someone is making careful, systematic observations with one or more specific questions in mind. There are at least two very general purposes of research. One is to portray accurately the basic characteristics of people, objects, and events in the world. The second is to develop statements about how different aspects of the world are related.

### Characterizing Observations

Suppose we were interested in the weight of newborn humans. We might attempt to weigh enough babies to develop an accurate picture of baby weights. How many are enough? Which babies should we weigh? What information would we have when we were finished? The details will be covered in Chapter 2, but clearly an accurate picture of baby weights would give us some idea of what a *typical* baby would weigh. Also, we would have some idea about the amount of *variability* there is among babies. Do they all tend to weigh about the same, or is there a fairly wide range of weights? As you will soon see, the weight of a typical baby could be expressed by a statistic—a number—that reflects what is referred to as the *central tendency* of all the weights we collected. The variability among weights could also be expressed by a number, one that we would want to be large if there was a lot of difference in weight among our babies and small if the babies did not differ much in weight. Certainly, we would hope that these numbers, our statistical indices of central tendency and variability, would accurately reflect the weights of babies in general, not just the few we have had an opportunity to observe and measure. We would then want to draw inferences about all babies based on our sample of information, and we would need to know how likely our inferences were to be correct. Is our picture of this aspect of the world likely to be accurate, or is there a good chance that it is distorted? An understanding of statistics opens the way to answering these questions.

### Characterizing Relationships

More often than not, the researcher wishes to go beyond characterizing isolated aspects of the world and to determine relationships among several phenomena. (Do larger parents tend to have larger babies? Do larger babies tend to be healthier than smaller ones? Will a large baby grow into a large adult?) In asking these more complex questions, further statistical indices will be required to tell us whether and to what degree two or more events are correlated.* It might be that the biggest infants invariably grow into the largest adults; at the other extreme, birth weight might be unrelated to size in adulthood. However, we might guess (and rightly) that the truth is somewhere in between. Birth sizes tell us something, but not all, about later physical stature. Where in this vast middle range does any particular relationship fall? Numerical measures (statistics) can be computed to answer each of these questions.

In the baby-weight example, emphasis was placed on simply finding a correlation between one characteristic (or, more technically, one *variable*)

*Thus, as the name implies, *correlation* refers to the co-, or joint, relationship between two or more characteristics.

and another; for example, we might ask about the correlation between a baby's weight and its health. Often, though, the social scientist wishes to go further and determine whether a *causal* relationship exists between one variable and another. Here, the mere presence of correlations is not sufficient.

Consider the following example. Educators have known for a long time that the amount of formal education a person receives and his or her later income are positively correlated; that is, people with more education tend to make more money. On the basis of this information, it is tempting to infer a causal relationship and assume that education actually causes an increase in later income through increasing the person's skills or, perhaps, making him or her more attractive to employers or better qualified for high-paying jobs. Although this type of cause-and-effect relationship may exist, the above information does not logically require or demand this conclusion. There are several competing explanations, or *rival hypotheses,* for the education-income correlation. Suppose, for example, that brighter people are simply more likely than less bright ones both to stay in school and to be clever enough to make more money. In such a case, education would not be the cause of higher income; rather, a third variable, intelligence, would be the cause of both education *and* income. Of course, this explanation is not necessarily the correct one either. The point is that when we simply measure existing characteristics and find an association (that is, a correlation), there are many possible explanations of what caused it. How are we to escape this dilemma? The answer lies in the *experimental method.*

*Testing for causal relationships: the experimental method.* The key to the experimental method is direct control of events by the investigator. Instead of merely looking for existing relationships, the experimentalist tries actually to produce them by manipulating the environment. Again, an example will be helpful.

Suppose you observed that the fur of minks tends to be longer in the winter than in the summer. You might, especially if you were thinking of going into the fur business, wonder what causes the difference. The most obvious possiblity is that cold weather stimulates the growth. But many other things change from summer to winter besides temperature, and one of these could be the controlling factor you are interested in. The animals' diet, for example, probably changes somewhat from season to season; the amount of daylight changes; the amount of humidity changes. Because all these factors vary according to the season of year, the only way to untangle them and to investigate their independent effects on the minks is by separately manipulating the various factors. This could be done in a series of simple experiments, each of which would test one hypothesis, or *potential* cause. For example, to determine whether temperature influences the length of fur, you could raise some minks under cool conditions (e.g., 50°F.) and some minks under warm conditions (e.g., 70°F.). In doing so, you would want to be sure that all other factors which might influence the length of the animals' fur were held *constant* (i.e., were the same for both groups of minks). One group should not tend to be older than the other, or in better general health, or from a different strain of minks with a genetic tendency for longer fur. In addition, you would want to be sure that the animals in the 50° temperature and those in the 70° temperature were given exactly the same diet, amount of light, humidity, and so on.

Suppose we have done everything we can to see that the only difference between the two groups of animals is the temperature of their cages. At the end of, say, three months we could measure the length of each animal's fur. We have now set up a true experiment, which can yield information about causal relations. The variable manipulated by the investigator (in this case, temperature) is called the *independent variable.* The measure of interest (in this case, length of fur) is called the *dependent variable.* Experiments are designed to determine whether the independent variable influences the dependent variable (e.g., does temperature affect the length of fur?).

Once we know the length of each animal's fur, we could calculate the average length for each group (an index of central tendency). Then we would face another problem. Regardless of the effects of temperature, these two numbers would almost certainly be different from one another. It would be surprising indeed if the average length of fur for one group was *exactly* the same as the average length of fur for the other group, even if the groups had not been in different temperatures and had been treated alike in every way. We would need a way of deciding if temperature really had any effect on fur length or if the difference in fur length between the two groups was just happenstance. In other words, are our results reliable? Would manipulating the temperature produce comparable results with other, similar samples of animals? One of the most important functions of statistical tests is to provide a basis for answering experimental questions such as these. To understand and use these tools, we must begin with the basic concepts of samples and populations.

# Samples and Populations

# 2

IMAGINE THAT YOU ARE AN ANIMAL PSYCHOLOGIST and want to know how many eggs are laid by nesting sea turtles on the island of Lith. With a lot of time and many helpers, you could mark each female as it comes onto the beach and assign a helper to watch her and later dig up the eggs from her nest and count them. In this way, you could observe all of the female sea turtles that nest on this particular beach and count the exact number of eggs each one deposited in her nest. If the helpers did not miss any females and the counting was accurate, the information obtained would be complete, or exhaustive. There would be little chance of any error in the picture yielded by such data.

Rarely can a study be carried out in this fashion. The time and the number of helpers required would be extremely costly, if not practically impossible. The number of eggs any one turtle produced might, for example, run into the hundreds. Then, too, there are ethical considerations. Disrupting the nests to count the eggs may prevent the eggs from hatching, and so we would want to disrupt the smallest number of nests that would still make it possible to obtain accurate information. Social scientists are continually faced with these same issues. Such circumstances merit a closer look.

Rather than try to observe all the turtles, the investigator might select a few of the females, count their eggs, take an average, and then generalize from these results to reach some conclusion about the typical number of eggs a sea turtle produces. Here there is room for error, and we will use our knowledge of statistics and research methodology as tools either to minimize this error or to fix it at an acceptable level.

## BASIC DEFINITIONS

### Populations

Research is almost always directed at characterizing and understanding a segment of the world, a *population*, on the basis of observing a smaller segment, a *sample*. As we use the term here, a population is the total collection of things under consideration. Populations are defined, not by nature, but by *rules of membership* invented by investigators. For example, all the female sea turtles of a given species that arrive on the beach during a specified period of time would make up the population in the preceding example. An investigator might be interested in all retarded children currently residing in a particular county, all nouns in the English language, all black-and-white movies produced in Sweden between 1956 and 1967, all yellow Plymouths, all medical clinics in Connecticut that handle more than fifty cases a day, or all left-handed people hopping on one foot anywhere in the world now.

A population may also be defined in terms of *potential* membership. An investigator might be interested in all U.S. adult males who might be given a particular birth control pill or all 120-day-old white rats of a particular strain who might be exposed to radiation.

### Samples

A *sample* is some subset of a population. A sample can be any size, as long as it consists of a number less than the total number of possible observations of a given type. Two very basic procedural questions arise in research: (1) *Which* cases should be sampled? (2) *How many* cases should be included in the sample?

*Representative samples: a goal.* The most accurate information about a population will come from a sample that is representative of the population from which it is selected. A *representative sample* tends to contain relevant characteristics in the same proportions as they exist in the population. Suppose that in our turtle example, 40 percent of the turtles are over three feet in diameter and 60 percent are under three feet. Then our sample, if it is to be representative, should also contain big and small turtles in approximately these same proportions, 40 percent and 60 percent. If 30 percent of the population is light green, 50 percent medium green, and 20 percent dark green, then these are the ideal proportions we would want in our sample, no matter what its size. This is true for any characteristic that you can think of, especially for any characteristic that you think might be related to the number of eggs a turtle produces.

A sample is said to be *biased* when it is not representative of the entire population to which an investigator wants to generalize. It is easy to see how a biased sample might give you an erroneous picture of the world. Assume that our sample of turtles happened to include 80 percent small female sea turtles and that, on the average, small sea turtles tend to lay fewer eggs than large sea turtles. In this case, we would underestimate the number of eggs laid by the entire population, which includes a larger percentage of big turtles. Any conclusions we reach about the average, or typical, sea turtle on the basis of such a distorted, or *nonrepresentative*, sample would probably not be accurate with regard to the population as a whole. In order to get an accurate picture of the population as a whole, all its characteristics must be represented in the sample in appropriate proportions. How can this be accomplished?

*Random samples.* A sample is random when (1) every member of the population has an equal chance of being selected to be in the sample and (2) the selection of any one member of the population does not influence the chances of selecting any other member. There are many ways to obtain random samples. One very simple way is to put the names or code numbers of all members of the population into a hat, shake them up, and without looking, draw out enough for your sample. This is usually the way winning lottery tickets are selected, and the procedure gives each ticket an equal chance of winning. An entirely analogous procedure involves the use of a table of random numbers. Appendix A (pages 211–214) contains such a table, and it will be helpful briefly to inspect a portion of it, shown in Table 2-1, now.

TABLE 2-1 Portion of random number table

| | | | | | |
|---|---|---|---|---|---|
| 39528 | 72784 | 82474 | 25593 | 48545 | 35247 |
| 81616 | 18711 | 53342 | 44276 | 75122 | 11724 |
| 07586 | 16120 | 82641 | 22820 | 92904 | 13141 |
| 90767 | 04235 | 13574 | 17200 | 69902 | 63742 |
| 40188 | 28193 | 29593 | 88627 | 94972 | 11598 |
| 34414 | 82157 | 86887 | 55087 | 19152 | 00023 |
| 63439 | 75363 | 44989 | 16822 | 36024 | 00867 |
| 67049 | 09070 | 93399 | 45547 | 94458 | 74284 |
| 79495 | 04146 | 52162 | 90286 | 54158 | 34243 |
| 91704 | 30552 | 04737 | 21031 | 75051 | 93029 |
| 94015 | 46874 | 32444 | 48277 | 59820 | 96163 |
| 74108 | 88222 | 88570 | 74015 | 25704 | 91035 |
| 62880 | 87873 | 95160 | 59221 | 22304 | 90314 |
| 11748 | 12102 | 80580 | 41867 | 17710 | 59621 |
| 17944 | 05600 | 60478 | 03343 | 25852 | 58905 |

The numbers in the table have been generated by a computer so that every digit is as likely to appear as every other. If you want to select a sample of 100 cases from a population of 500, you would enter the table at any point (e.g., the upper-right-hand corner) and then read three-digit numbers until you had found 100 numbers between the values of 001 and 500. Of course, any numbers you encountered that were larger than 500 would be ignored. In practice, samples are usually selected *without replacement*, which means that a number (or individual) may be selected only once for any one sample. For example, using Table 2-1, our first ten numbers would be: 247, 141, 23, 284, 243, 29, 163, 35, 314, 122. Note that 247 is the last three digits of the upper right-hand entry (35247), that 724 is skipped because it is larger than 500 so that 141 is the next number selected, and that 023 is equivalent to 23. When the end of a particular column is reached, one can move arbitrarily to either the top or bottom of the next column. In this example, we moved to the top and thus the last number in our sample is 122.

This procedure, it can be shown, will tend to yield a representative sample. For example, if there were 100 turtles in our population, 40 large and 60 small, the probability of selecting a large turtle to be in our sample on any given drawing would be 0.4, and the probability of selecting a small one would be 0.6. Given this, the characteristics of large and small should be represented in our sample in the same proportions that they exist in the population: 4 to 6. Of course, even a carefully drawn random sample will rarely, if ever, exactly mirror the characteristics of the population, but it will *tend* to do so in the long run.

By drawing our samples randomly, we hope to make them representative with respect to all relevant characteristics of the population. That is, random sampling does not eliminate all possibility of error. But it does guard against any *systematic biases* slipping into the selection of a sample.

To illustrate the idea of a biased sample, imagine that a research team is trying to obtain opinions from a sample of employees of a certain company on a range of questions having to do with sexual attitudes. They decided to select a sample of 25 people and simply went to the company cafeteria one noontime and asked 25 people to fill out a questionnaire. They then generalized the results to all the employees of the company. What is wrong with this procedure of selecting a sample?

Two of the things that might occur are:

1. All employees may not eat lunch in the cafeteria. It may be that those who do eat in the cafeteria tend to be different from those who do not. Perhaps, for example, the higher-paid executives are the ones who go out for lunch; if they also have sexual attitudes that are different from those of employees who eat in, then their views will tend to be overlooked, and the sample will be biased.
2. In selecting the sample in the cafeteria, perhaps the investigator would tend to ask people who looked nice or especially cooperative or who did not seem

too busy to fill out the questionnaire. This might especially be the case with a questionnaire on a topic like sex. If those people who look nice or cooperative or who are not engaged in conversation with others have different attitudes than those who look less cooperative or very busy, the sample will be biased and will not accurately reflect or represent the views of the population as a whole.

If generalizations are to be made to an entire population, then all members must have had an equal chance to be selected. A random procedure, such as numbering the names in a company phone book and then selecting the sample with a table of random numbers, guarantees that such things as eating habits or appearance do not influence the selection of the sample. Random sampling provides the best defense against the possibility of a systematic bias operating during the selection of a sample.

But another question arises: How many cases should be selected? The answer to this one is somewhat more difficult to explain now, but the factors that influence this decision will become clearer to you as you read this book. Such things as tradition (what other investigators in the same field do) enter into the decision. However, probably the thing that is most influential for deciding on a sample size is what the investigator knows or expects about the variability in the population being studied. If everybody were exactly alike, a sample of only one case would always be fully representative of the population. It is because people, objects, and events differ from one another in so many interesting ways that we need samples larger than one; how much larger often is related to how great we expect the differences to be.

Suppose, for example, that a team of scientists from another planet landed on Earth and met one man. They observed him carefully and noted he (1) had two eyes, (2) had ten fingers, (3) reported having one sister, and (4) could play the "Star-Spangled Banner" on a ukelele with his toes. *We* know (but our extraterrestrial guests would not) that this sample of one almost fully represents the population (of all people) on some characteristics (e.g., having two eyes), represents a fairly sizable portion on others (e.g., having one sister), and is almost unique in yet other characteristics, such as his musical ability. Without this knowledge, our imaginary research team might mistakenly guess that all humans can play the ukelele with their toes.

## CHARACTERIZING OBSERVATIONS

Assume you are a political reporter and you want to get some idea of the level of interest in the abortion issue. As part of your work, you decide to sample members of the U.S. House of Representatives and find out how many letters mentioning this issue they received last week. You might put the names of all the representatives in a box and draw out 12 in order to obtain a random sample. For reasons similar to those discussed in the last section, this procedure would obviously be preferable to calling up the 12 representatives with whom you are most friendly.

Table 2-2 presents the data from this hypothetical study, employing the general form we will use throughout this book. The $X$ at the head of the column is a general way of designating the *value* or score of the characteristic or performance observed for each subject. In this case, it is the number of relevant letters each subject reported receiving. (The terms *score*, *observation*, *measure*, and sometimes, *datum* are often used interchangeably.)

TABLE 2-2 Number of letters per week mentioning abortion reported by a random sample of 12 U.S. Representatives*

| Number of letters ($X$) |
|---|
| 2 |
| 4 |
| 4 |
| 1 |
| 11 |
| 4 |
| 0 |
| 1 |
| 0 |
| 7 |
| 4 |
| 1 |

*These data and all the illustrations that follow are based on hypothetical rather than actual findings.

Now that you have your sample of observations, you want to get the clearest possible picture of the information yielded by the sample. One way to clarify the results is to draw a *frequency distribution*, such as that shown in Figure 2-1. The horizontal axis shows possible values of the scores (number of letters) and the vertical axis shows the number of observations of each value (number of representatives receiving a particular number of letters). Each asterisk (*) in the figure indicates one representative who received the number of letters indicated directly below it on the horizontal axis. Thus, inasmuch as two representatives received no letters at all, there are two asterisks in the 0 column. When the data from Table 2-2 have been plotted in this way (in Figure 2-1), the form of the distribution of scores is more apparent, and it is quite easy to see which scores appear typical (e.g., 4 letters) and which seem more unusual or deviant for this particular sample (e.g., 11 letters).

There are at least two major aspects of any set of scores that you will

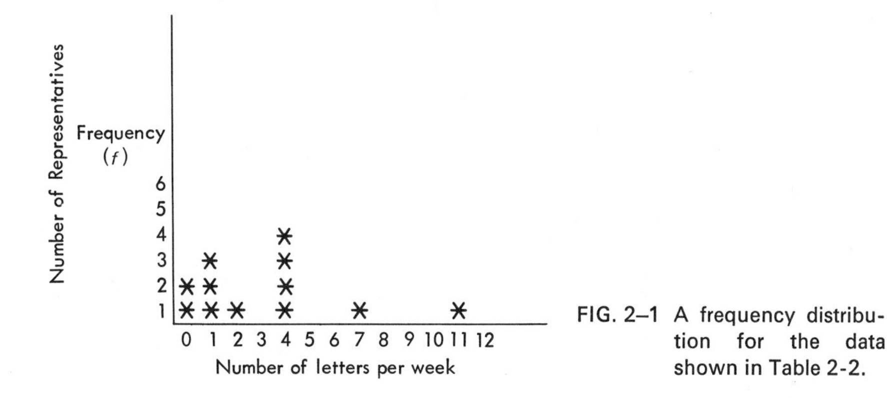


FIG. 2–1 A frequency distribution for the data shown in Table 2-2.

want to summarize and communicate to other investigators. (1) You will want to describe *where* the scores tend to be, that is, whether they are high or low. For example, did representatives in general receive a lot of letters mentioning abortion or only a few? (2) You will want to note whether the scores tend to be alike or whether there is a lot of *variability* among them. Did most representatives receive about the same number of letters, or did some receive a great many and others only a few?

Figure 2-2 further illustrates these two concepts. The data are four samples of reading-comprehension scores, obtained from four different populations of students. Note that *b* and *d* are similar in variability (both containing a number of low and high scores); *a* and *c* are also similar to one another, but in each of these distributions, the variability is small and all the scores are nearly alike. Distributions *b* and *d* are more variable than *a* and *c*. Note, however, that *a* and *c* are different from one another in another dimension, central tendency. The typical score in distribution *a* is 5, whereas the typical score in *c* is 8. In a similar way, *b* differs from *d* in central tendency.

### Indices of Central Tendency or Location

There are three common measures of central tendency.

*Mode.* The mode is that score which occurs most frequently in a distribution. In Figure 2-2, the mode for distribution *a* is 5. A score of 5 was obtained by more people than any other score on the test.

*Median.* The median is that score above and below which 50 percent of the observations fall. It is usually found by ranking the scores from highest to lowest and then counting halfway down the list. If there are an odd number of scores, the median is the middle score. If there are an even number of scores, the median is halfway between the two middle scores. The median of the scores in distribution *a*, Figure 2-2, is 5.

*Mean.* The arithmetic mean is the most familiar and probably the most useful measure of central tendency. The mean (commonly known as the

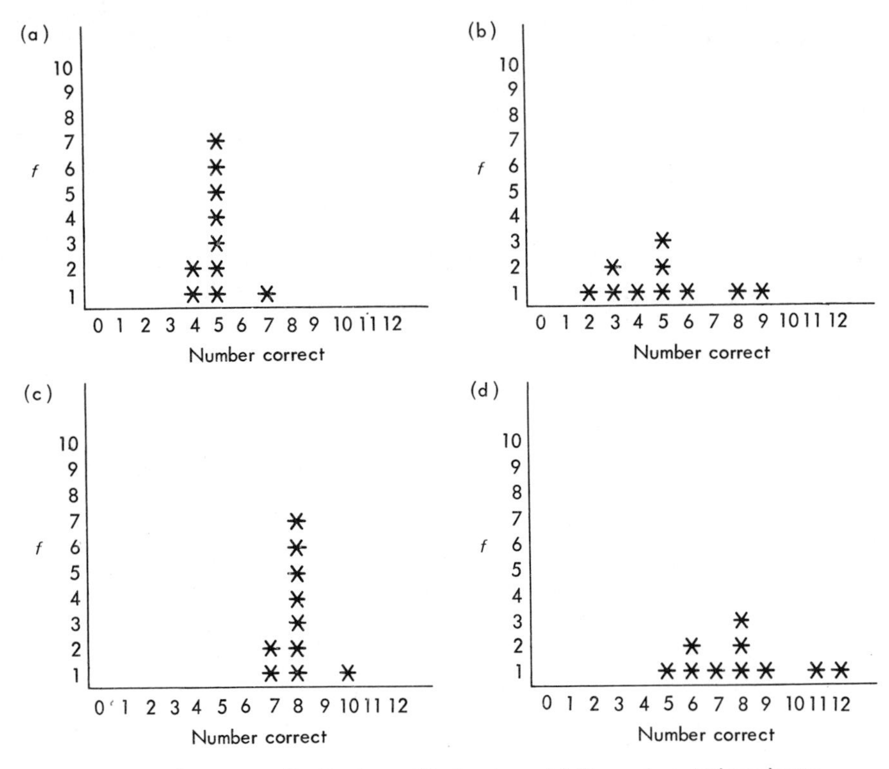


FIG. 2–2 Frequency distributions differing in variability and central tendency.

*average*) is found by summing all the scores and dividing by the number of scores. These operations are indicated in the following formula:

$$\bar{X} = \frac{\sum X}{N}$$

Here $\bar{X}$ is the mean, $X$ refers to each score, and $N$ is the total number of scores. $\sum$ is the symbol for summation. Thus, in words, the formula reads: The mean equals the sum of the $X$s divided by $N$.

The mean of sample *a* in Figure 2-2 is identified by a subscript. Thus

$$\bar{X}_a = \frac{\sum X}{N}$$

$$= \frac{4 + 4 + 5 + 5 + 5 + 5 + 5 + 5 + 5 + 7}{10}$$

$$= \frac{50}{10}$$

$$= 5.00$$

The mean of distribution *b* is

$$\bar{X}_b = \frac{\sum X}{N}$$

$$= \frac{2 + 3 + 3 + 4 + 5 + 5 + 5 + 6 + 8 + 9}{10}$$

$$= \frac{50}{10}$$

$$= 5.00$$

Notice that the means of both distributions are the same. By computation, you should now be able to determine $\bar{X}_c$ and $\bar{X}_d$.*

Note that in each of the four distributions in Figure 2-2, the mean, median, and mode all have the same value. This is not always the case; in Figure 2-1 (page 13), the mode of the distribution is 4, the median is 3, and the mean is 3.25. Which of these values is the "best" index of central tendency? Certainly, you could influence the picture somewhat, depending on which index you reported as the typical number of letters received by representatives. All are appropriate, so we must ask which best represents the scores in the light of the purpose to which the information is being put. If you are manufacturing shirts, you would probably be more interested in the modal size of people than in the mean size (especially considering that no one may actually *be* the mean size). For many other purposes, however, the mean is a particularly useful and informative measure of location and is the most commonly used in studies involving statistical inference. Unless otherwise specified, you may assume throughout this book that average is equivalent to the mean. Usually, a sample mean ($\bar{X}$) is obtained for the purpose of estimating a population mean, designated by the Greek letter *mu* ($\mu$).

### Indices of Variability

Just as the mean is an index of location, we would also like to have an index of variability. This index should, for example, capture the difference between distributions *a* and *b* in Figure 2-2.

*Range.* One very simple index of variability is the *range.* The range is the difference between the highest and lowest scores in the distribution. The range for sample *a* in Figure 2-2 is $7 - 4 = 3$; the range for *b* is $9 - 2 = 7$. As you can see, this is an easily calculated index of the amount of variation in the scores. Probably the most severe problem with using the range as our exclusive index of variability is that it includes so little of the information provided by our data. Suppose, for example, we had a set of 900 scores, one

*In both cases, the mean is 8.00.

of which had the value of 1,000, while each of the others had the value of 60. Although there is really very little variability, the computed value of the range would be 940! Therefore, except to get a quick look at what happened in a study before you have time for more detailed analyses, the range is rarely used.

*Sum of squares.* One very useful way of expressing variability, which is based on every score, depends on how much the scores in a distribution deviate from the typical score. Are they generally close to the average score or not? This concept of variability is reflected in an index called the *sum of squares* (a shortened form of the *sum of squared deviations from the mean*). The formula for the sum of squares, abbreviated $SS$, is

$$SS = \sum (X - \bar{X})^2$$

First, look at the quantity within the parentheses:

$$X - \bar{X}$$

This indicates that the mean of the distribution ($\bar{X}$) is to be subtracted from each score within that distribution. For example, take the subject in sample *a* (Figure 2-2) who has a score of 7. Then

$$X - \bar{X} = 7 - 5 = 2$$

This subject deviates from the most typical score (the mean) by 2 units. A subject with a score of 4 deviates from the mean by $-1$ unit (the minus sign indicates the score is below the mean). Refer now to the left side of Table 2-3 (we will discuss the side of the table labeled "computational formula" later). In this table, you will find each subject's *deviation score* ($X - \bar{X}$). Now compare the deviation scores for distributions *a* and *b*. You should immediately notice that, ignoring the sign of the scores, they tend to be greater in distribution *b* than in distribution *a*. This reflects the fact that the scores tend to be rather widely dispersed in *b*; whereas they tend to be more tightly clustered together in *a*.

Because the fact that deviation scores are signed (i.e., $+$ or $-$) is mathematically inconvenient, it is desirable to eliminate the signs in some manner other than just ignoring them. Squaring each deviation score does just this while preserving the important relationship of its relative value compared with other scores. Now look at the column in Table 2-3 in which each deviation score has been squared. For the first subject

$$(X - \bar{X})^2 = (4 - 5)^2 = (-1)^2 = 1$$

TABLE 2-3 Reading scores from samples *a* and *b* from Figure 2-2

| | | Defining formula | | Computational formula | |
|---|---|---|---|---|---|
| *Sample a* | $X$ | $(X - \bar{X})$ | $(X - \bar{X})^2$ | $X$ | $X^2$ |
| | 4 | $(4 - 5) = -1$ | $(-1)^2 = 1$ | 4 | 16 |
| | 4 | $(4 - 5) = -1$ | $(-1)^2 = 1$ | 4 | 16 |
| | 5 | $(5 - 5) = 0$ | $(0)^2 = 0$ | 5 | 25 |
| | 5 | $(5 - 5) = 0$ | $(0)^2 = 0$ | 5 | 25 |
| | 5 | $(5 - 5) = 0$ | $(0)^2 = 0$ | 5 | 25 |
| | 5 | $(5 - 5) = 0$ | $(0)^2 = 0$ | 5 | 25 |
| | 5 | $(5 - 5) = 0$ | $(0)^2 = 0$ | 5 | 25 |
| | 5 | $(5 - 5) = 0$ | $(0)^2 = 0$ | 5 | 25 |
| | 5 | $(5 - 5) = 0$ | $(0)^2 = 0$ | 5 | 25 |
| | 7 | $(7 - 5) = 2$ | $(2)^2 = 4$ | 7 | 49 |
| $\Sigma$ | 50 | | 6 | 50 | 256 |
| *Sample b* | $X$ | $(X - \bar{X})$ | $(X - \bar{X})^2$ | | |
| | 2 | $(2 - 5) = -3$ | $(-3)^2 = 9$ | | |
| | 3 | $(3 - 5) = -2$ | $(-2)^2 = 4$ | | |
| | 3 | $(3 - 5) = -2$ | $(-2)^2 = 4$ | | |
| | 4 | $(4 - 5) = -1$ | $(-1)^2 = 1$ | | |
| | 5 | $(5 - 5) = 0$ | $(0)^2 = 0$ | | |
| | 5 | $(5 - 5) = 0$ | $(0)^2 = 0$ | | |
| | 5 | $(5 - 5) = 0$ | $(0)^2 = 0$ | | |
| | 6 | $(6 - 5) = 1$ | $(1)^2 = 1$ | | |
| | 8 | $(8 - 5) = 3$ | $(3)^2 = 9$ | | |
| | 9 | $(9 - 5) = 4$ | $(4)^2 = 16$ | | |
| $\Sigma$ | 50 | | 44 | | |

Finally, of course, the symbol $\Sigma$ again refers to summation and requires that we add the squared individual deviation scores; their total, or sum, characterizes the entire distribution.

To illustrate the computation, compare distributions *a* and *b* and look at the sums of the squared deviation scores.

$$SS_a = \Sigma (X - \bar{X})^2 = 1 + 1 + 0 + 0 + 0 + 0 + 0 + 0 + 0 + 4 = 6$$

and

$$SS_b = \Sigma (X - \bar{X})^2 = 9 + 4 + 4 + 1 + 0 + 0 + 0 + 1 + 9 + 16 = 44$$

As you can see, the sum of squares reflects the greater variability in the second distribution.

*Sample variance.* The use of the sum of squares as a measurement of variability sometimes has a disadvantage: its size depends on the number of scores. For example, if you were comparing two distributions that did not have an equal number of scores, the distribution with less variability might have the larger $SS$ simply because more scores entered into the total. Therefore, the $SS$ is often divided by the number of scores in order to obtain an *average* sum of squares, called the *sample variance*

$$\frac{SS}{N}$$

The sample variance retains the most important characteristic of $SS$: the greater the variability or dispersion of the observations around the mean, the larger $SS/N$ will be. For distribution *a*

$$\frac{SS}{N} = \text{sample variance} = \frac{6}{10} = 0.60$$

Verify that the sample variance for distribution *b* is 4.40.

*Variance estimate.* Usually, we compute a sample variance not only to describe the variability in the sample but also to estimate the population variance. The population variance is designated $\sigma^2$ (sigma squared). Remember, our goal is to make inferences about populations on the basis of a subset of cases, the sample. A sample is, by definition, smaller than the population from which it was obtained. The smaller the sample, the less likely it is to include extreme cases from the population. Extreme or very deviant cases are very often the scores that are the least frequent in the population and therefore have less chance of being observed relative to less deviant scores. Extreme cases do, of course, contribute to and increase the overall population variance. However, if they are not included in the sample, they will not contribute to our *estimate* of the population variance, and the estimate will be too small. An estimate of the population variance based on a sample variance thus tends to be lower than it should be, and the degree to which the sample variance underestimates the population variance tends to be greater with small samples than with large ones.

Therefore, the statistical procedures that we will be describing later all use a modified version of the above formula, called the *variance estimate* or $s^2$.

$$s^2 = \text{variance estimate} = \frac{SS}{N - 1}$$

As you can see, $N - 1$ is used in the denominator, rather than $N$. This produces a correction in the value of the sample variance that is related to sample size and that brings it closer to the value of the population variance. If you think about it, as the sample size ($N$) gets larger, subtracting 1 makes less difference in the computed value of the variance; but for small samples, which are the main problem, it makes more of a difference.

The variance* for distribution *a* is

$$s_a^2 = \frac{SS}{N-1} = \frac{6}{9} = 0.67\dagger$$

Verify that $s^2$ for distribution *b* is 4.89.

Note that the computed value of the variance, like that of *SS*, is larger in the distribution with the greater spread or variability among subjects.

*Standard deviation.* Another frequently used index of variability is the *standard deviation,* abbreviated *s*. This is used to estimate the standard deviation of the population, designated $\sigma$ (sigma). The standard deviation is simply the square root of the variance.

$$\text{Standard deviation} = \sqrt{s^2} = s$$

The standard deviations of distributions *a* and *b* are:

$$s_a = \sqrt{s_a^2} = \sqrt{0.67} = 0.82$$
$$s_b = \sqrt{s_b^2} = \sqrt{4.89} = 2.21$$

Again, you will notice that the distribution with the greater variability among subjects yields the larger computed value of the variability index (in this case, the standard deviation).

Obviously, *SS*, $s^2$, and *s* are closely related indices of variability. They all reflect or express the extent to which individual scores deviate from the mean score of their distributions. As you work through this book, you will get an idea of the different situations in which one or the other of these variability measures is used. Right now, they all should have approximately the same meaning to you. The greater variability in samples *b* and *d* indicates that the reading abilities of these subjects is more diverse than that of the subjects in *a* or *c*. (This fact might help us decide which program would be best in each of the populations.)

### Computational Formula for the Sum of Squares

The following formula, introduced previously (page 16), emphasizes the meaning of variability:

$$SS = \sum (X - \bar{X})^2$$

*Because we will always use the information from a sample to estimate the population variance, the shorter term *variance* will hereafter refer to variance estimate.

†In this book most calculations have been rounded to two decimal places at each step. If you carry computations out further, or use a "pocket calculator," your final answer may differ from ours by 1–2 percent.

However, an equivalent formula that will simplify your calculations a great deal under many circumstances is

$$SS = \sum X^2 - \frac{(\sum X)^2}{N}$$

A proof that $\sum (X - \bar{X})^2 = \sum X^2 - (\sum X)^2/N$ can be found in Appendix B. Right now, you should memorize both formulas and get in the habit of using the second (computational) formula whenever you are going to compute the value of the $SS$, $s^2$, or $s$. The right side of Table 2-3 shows you how to set up the data for using the computational formula. The computations are illustrated below:

$$\begin{aligned} SS &= \sum X^2 - \frac{(\sum X)^2}{N} \\ &= (16 + 16 + 25 + 25 + 25 + 25 + 25 + 25 + 25 + 49) \\ &\quad - \frac{(4 + 4 + 5 + 5 + 5 + 5 + 5 + 5 + 5 + 7)^2}{10} \\ &= 256 - \frac{(50)^2}{10} \end{aligned}$$

As you can see, the values for $\sum X^2$ and $(\sum X)$ can be taken directly from Table 2-3.

$$\begin{aligned} &= 256 - \frac{2500}{10} \\ &= 256 - 250 \\ &= 6.00 \end{aligned}$$

## Standard Scores

It is a scientific truism that an observation is only meaningful in relation to other observations. Knowledge of the *relative* status of a score is central to interpreting its meaning. For example, if we tell you that Bill Smith is 6 feet tall, you understand the statement partly in terms of your knowledge and/or assumptions about the underlying distribution of heights in the population. You have some idea of the range of heights among adult males and some notion of the relative frequency of these heights. That is how you know that Bill Smith is tall as opposed to short. If we tell you that the distance from the floor to the roof in a camper we are thinking of buying is 5 1/2 feet, you not only know that Bill Smith will not be able to stand up in it, but you also probably have some intuitive feel for the percentage of adult people who will not be able to stand up in the camper.

A standard score converts raw scores into a form that makes these intuitive judgments more precise by indicating the relative status of a particular score in a distribution. For example, if we tell you that Bill Smith has a score of 30 on a test of hostility, you would probably ask, "Out of how many possible?" If we tell you the maximum possible hostility score is 60, you might say Bill was not very hostile. On the other hand, if we tell you that the mean score on this test was 15, you might think Bill was quite hostile, even unusually hostile. Now, if we add the standard deviation of the test scores to the information already given you, you will have some idea of just how unusual Bill is. If the standard deviation of the distribution is 5, the scores will tend to be closer to the mean, on the average, than if the standard deviation is 10. Therefore, Bill Smith's score of 30 is more unique in a distribution of scores with $\bar{X} = 15$ and $s = 5$ than in a distribution of scores with $\bar{X} = 15$ and $s = 10$. A standard score summarizes all this information in a single value. The concept is illustrated in Figure 2-3, in which two possible distributions of hostility scores are shown. (You might also notice that we have shown both distributions as roughly "bell-shaped." In fact, many distributions of social science data do have this general form, a matter about which we will have more to say in Chapter 4.)

In general, a raw score can be converted into a standard score that expresses its deviation from the mean in standard-deviation units, according to the formula

$$\text{Standard score} = \frac{X - \bar{X}}{s}$$

If we compute Bill Smith's standard score for each of the two distributions described

$$\text{Standard score}_a = \frac{30 - 15}{5} \qquad \text{Standard score}_b = \frac{30 - 15}{10}$$

it becomes apparent that Bill's score is 3 units away from the mean in the first distribution but only 1 1/2 units away in the second distribution. We would say in the first case that he is 3 standard-deviation units, or simply 3 standard deviations, from the mean and in the second that he is 1 1/2 standard deviations from the mean.

In sum, then, raw scores are transformed into standard scores when you want to emphasize the relative location or status of a particular score in a distribution or when you want to compare a person's relative standing in two different distributions.

### Example 1

A score of 4 has different meanings in distributions *a* and *b* in Figure 2-2. We have already computed the means and standard deviations of these two

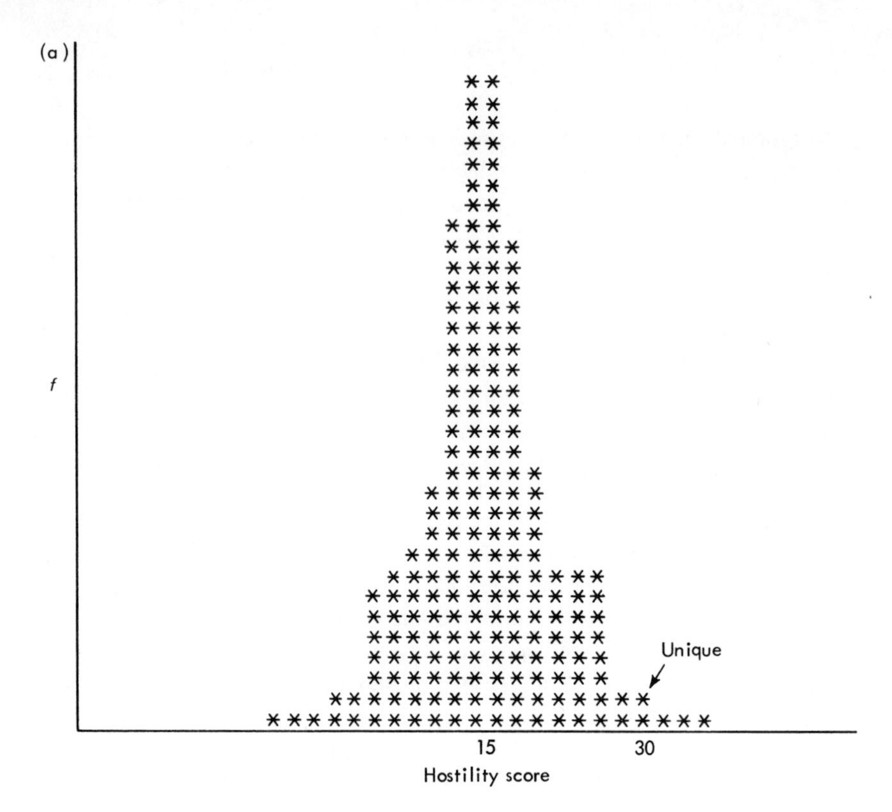


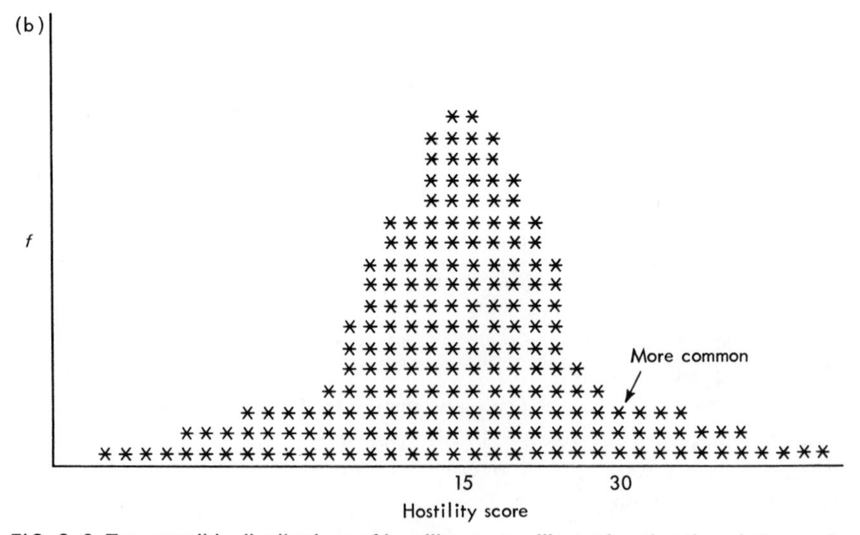


FIG. 2–3 Two possible distributions of hostility scores, illustrating that the relative position of a score in its distribution is as important as the actual value of the score itself. Bill's score (30), designated by the arrows, is the same in both distributions, but it is more unique (i.e., Bill is *relatively* more hostile) in distribution *a* than in distribution *b*.

distributions (page 19). The standard scores corresponding to a score of 4 in each distribution are

$$\text{Standard score}_a = \frac{X - \bar{X}}{s} = \frac{4 - 5}{0.82} = \frac{-1}{0.82} = -1.22$$

$$\text{Standard score}_b = \frac{X - \bar{X}}{s} = \frac{4 - 5}{2.21} = \frac{-1}{2.21} = -0.45$$

These standard scores indicate that a score of 4 is below the mean in both distributions (because the values of the standard scores are negative). However, a score of 4 is more deviant in distribution *a* than it is in distribution *b*.

EXAMPLE 2

Compare a score of 7 in distribution *a* with a score of 9 in distribution *b*. Because both distributions have a mean of 5, it might seem at first that 9 is the more deviant score. Now, compute the standard scores in each case.

$$\text{Standard score}_a = \frac{7 - 5}{0.82} = 2.44$$

$$\text{Standard score}_b = \frac{9 - 5}{2.21} = 1.81$$

Thus, a score of 7 is more deviant in sample *a* than a score of 9 is in sample *b*. The former is more than 2 standard-deviation units above the mean, and the latter is less than 2 above. Verify for yourself that even a score of 10 in distribution *b* would not be as deviant as a score of 7 is in *a*.

The basic point of these examples is that a raw score by itself often does not convey much meaningful information. A great deal is added by knowing the mean and the standard deviation of the distribution from which any particular score comes.

## Practice Problems

### *A. Look Out for Falling Garbage?*

A recent newspaper article reported that New York City garbagemen suffer three times as many injuries as coal miners do. In addition, the garbagemen suffer heart disease at a rate almost twice that for other men in similar age categories and also show a higher rate of arthritis. The article concluded that prolonged exposure to garbage is detrimental to your health.

Comment on this conclusion.

## B. *Pain Other than Headache?*

"In a recent study at a leading hospital, patients suffering from pain other than headache rated Panacea-X as more effective than the other leading brand of headache remedy."

Comment on this advertisement.

## C. *Lawbreakers on the Street Corner*

You are interested in automobile driving patterns in a certain city. You decide to watch 20 intersections for one hour each and to record the number of traffic violations occurring at each intersection (e.g., failure to yield the right-of-way, running a red light, and so on). Describe an appropriate procedure for randomly sampling 20 intersections.

## D. *Talk of the Town*

You are studying social interaction in a residential home for older citizens. You record the number of separate times each person in a random sample of residents participates in the dinner conversation during a 30-minute observation period. The data are given below.

| Subject | $X$ |
|---|---|
| 1 | 9 |
| 2 | 0 |
| 3 | 4 |
| 4 | 1 |
| 5 | 4 |
| 6 | 0 |
| 7 | 0 |
| 8 | 2 |
| 9 | 0 |
| 10 | 15 |
| 11 | 0 |
| 12 | 1 |

1. Draw a frequency graph for these data.
2. Calculate: mode, median, and mean.
3. Calculate: range, sum of squares, variance, and standard deviation.
4. What is the standard score for Subject 8? For Subject 10? Which of these two people is least characteristic of the group?

## E. *Classy Kids*

You are interested in the extent to which children are aware of status symbols. You obtain a sample of children by randomly selecting 10 names from the membership list of cub scouts in Posh County, and you get permission to interview each child. The interview consists of 10 questions, each requiring the child to choose between a high-status and a low-status item (e.g., "Would you rather have your parents buy a Ford or a Lincoln?"). The data are given below.

| Number of high-status choices ($X$) |
|---|
| 3 |
| 2 |
| 1 |
| 2 |
| 1 |
| 2 |
| 3 |
| 2 |
| 1 |
| 2 |

1. Compute the mean and the variance of this sample.
2. The sample mean and variance are estimates of the mean and variance of what population?

## *F. Neanderthal Error?*

*Scientific World* magazine has just reported a remarkable new anthropological finding: the skeletal remains of a creature similar to Neanderthal man, but one that stood less than 4 feet high.

Is this skeleton clear evidence of a new missing link, as the report suggests?

## *G. The Numbers Game*

Given a distribution of scores, from what part of the distribution could you remove cases and thereby increase the standard deviation?

# The Sample Mean: What Does It Mean?

# 3

LAST YEAR, 100 PEOPLE IN A PARTICULAR COUNTY in California bought one of the ten most expensive American-made automobiles. Assume you are conducting a study of the shopping habits of buyers of expensive cars in this county. You decide to select a random sample of 10 buyers (from the entire population of 100 buyers) and to interview them at length. One of the questions you ask them is how many other cars they examined prior to making their choice.

Of course, it is not possible for you actually to do this experiment now, but it can be simulated so that we can illustrate some important features. Table 3-1 shows the scores (number of other cars considered before purchase) for all the 100 people in our population and a frequency graph of the same information. From the numerical list of scores, you should randomly select a sample of 10, using the Table of Random Numbers. (The table appears in Appendix A; detailed instructions for its use appear on pages 9–10).

Now, compute the mean of the sample you have just drawn. Because the scores in the population range from 0 to 9, your mean should also fall in this interval. If, by chance, you selected 10 scores that were all 0, then you would have a mean of 0. Similarly, if you selected 10 scores that were all 9,

```
          **
          **
          **
          **
          **
          **
          **
          **
         ****
         ****
   f     ****
         ****
         ****
         ****
        ******
        ******
        ******
        ******
       ********
       ********
      **********
      **********
      0123456789
```

TABLE 3-1 Hypothetical population of the 100 expensive-car buyers in County A, showing the number of cars each buyer examined before making a purchase

| *Population A* | | | | | | | | | |
|---|---|---|---|---|---|---|---|---|---|
| 4 | 3 | 7 | 6 | 4 | 3 | 5 | 3 | 5 | 3 |
| 5 | 4 | 3 | 4 | 5 | 6 | 0 | 6 | 6 | 8 |
| 3 | 6 | 3 | 5 | 4 | 2 | 1 | 5 | 2 | 7 |
| 6 | 6 | 6 | 4 | 4 | 4 | 9 | 4 | 4 | 4 |
| 5 | 2 | 1 | 7 | 5 | 1 | 3 | 2 | 2 | 5 |
| 4 | 2 | 4 | 5 | 2 | 7 | 7 | 6 | 6 | 4 |
| 1 | 5 | 2 | 7 | 5 | 8 | 5 | 4 | 4 | 3 |
| 5 | 4 | 6 | 8 | 6 | 5 | 5 | 7 | 4 | 6 |
| 7 | 3 | 3 | 4 | 9 | 5 | 5 | 5 | 4 | 6 |
| 5 | 3 | 8 | 0 | 5 | 4 | 3 | 4 | 5 | 3 |

then you would have a mean of 9. Probably, your sample mean is somewhere between 0 and 9.

Suppose, though, you had selected a different random sample. How might it differ from the first? You can answer, in part, by actually drawing randomly a second sample of 10 observations from Table 3-1.* After doing so, compute the mean of the second sample. Chances are it does not equal the mean of the first sample. Which mean is a better estimate of the value of the population mean ($\mu$)? Why are the two means unequal?

The two sample means are equally good estimates of the population mean. There is no obvious reason to prefer one to the other. However, it would not be surprising if each differed somewhat from the population mean by chance. The fact that a sample mean usually differs from the mean of the population from which it comes is called *sampling error*. This term reflects the fact that the sample mean may be in error as an estimate of the true mean of the population. Similarly, two different samples from the same population may be in error in different directions or by different amounts, or both. Our job becomes clear: to estimate the usual amount of sampling error that will result from samples of various sizes.

## SAMPLING DISTRIBUTION OF THE MEAN

Imagine that we had drawn many random samples of 10 scores from Table 3-1 and computed the mean of each. We could then determine the distribu-

*You should draw this second sample from the full set of 100 numbers in Table 3-1, even if this means that one or more of your selections appears in both samples.

tion of these sample means, technically referred to as the *sampling distribution of the mean*, as has been done in Figure 3-1. Now we can actually see the amount and range of sampling error that is likely to occur when $N = 10$.

What would happen to the distribution of the sampling error, though, if we had taken samples of size $N = 30$, instead of $N = 10$? Figure 3-2 shows a distribution of means computed from random samples of 30 numbers drawn from the population of car buyers in Table 3-1. As you can see, there is obviously less variability among the individual means computed from the larger sample. However, the *overall mean* of each sampling distribution tends to be the same in both cases ($N = 10$ and $N = 30$).

But now another question arises. What happens to the distribution of sample means when we sample from a population with greater *variability* than the one in Table 3-1?

Table 3-2 shows such a population, which you might wish to think of as all the buyers of expensive cars in another county, County B. The scores of buyers from County B are more variable than are those of buyers from County A. They differ more among themselves in terms of the number of

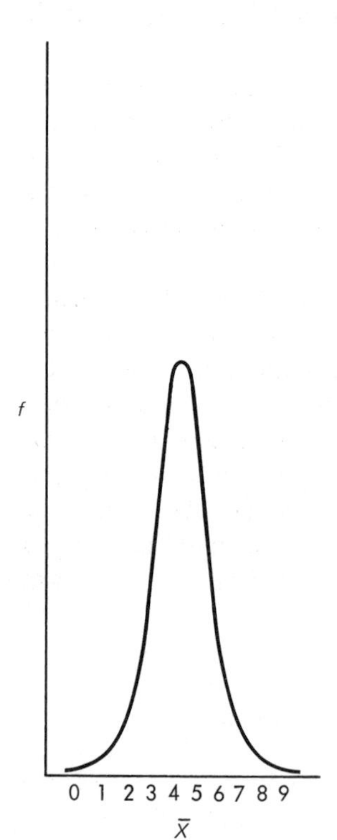


FIG. 3–1 A sampling distribution of the mean ($N = 10$) for the population of scores shown in Table 3-1.

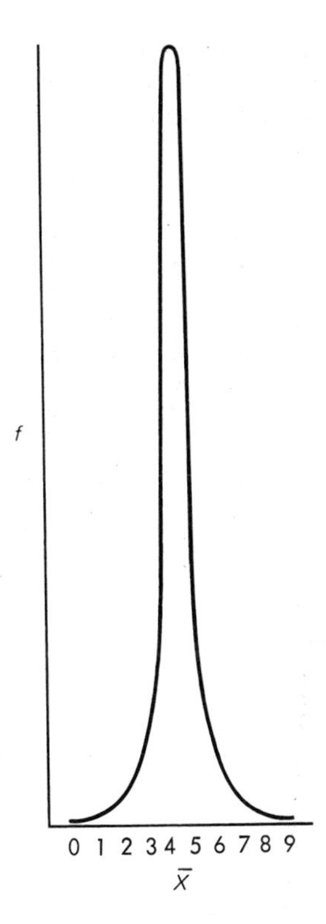


FIG. 3–2 A sampling distribution of the mean (N = 30) for the population of scores shown in Table 3-1.

cars they looked at before buying. From this population, we have drawn numerous samples with $N = 10$ and $N = 30$, computed the mean of each, and then plotted the means. Figure 3-3 displays the sampling distributions of the mean for samples of $N = 10$ and $N = 30$, respectively, for the County B set of scores. As you can see by comparing these two sampling distributions of the mean with those in Figures 3-1 and 3-2, greater variability in the original set of scores will produce greater variability in the values of the sample means. Also notice that, again, a smaller sample yields more variability in the sampling error.

Finally, what happens to the sampling distribution of the mean when we sample from an original population of scores with a shape that is very different from the ones we have been considering? Table 3-3 shows the scores from buyers in County C. As you can see, here the population has a flat shape, in contrast with the rather peaked shapes of Populations A and B. Figure 3-4 illustrates sampling distributions of the mean for sample sizes of $N = 10$ and $N = 30$, respectively, from Population C.

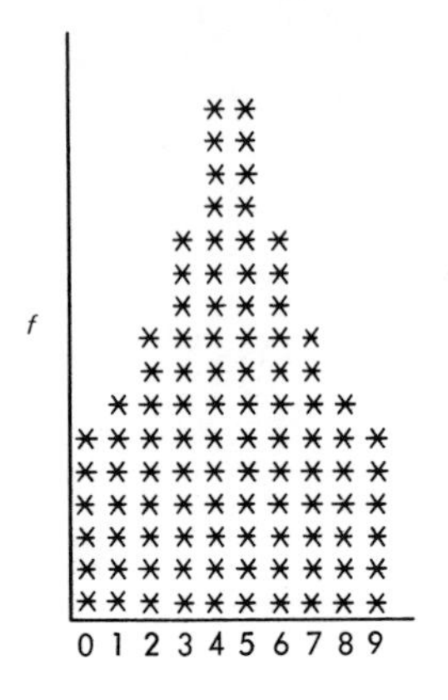


TABLE 3-2 Hypothetical population of the 100 expensive-car buyers in County B, showing the number of cars each buyer examined before making a purchase

| *Population B* | | | | | | | | | |
|---|---|---|---|---|---|---|---|---|---|
| 4 | 5 | 1 | 6 | 5 | 9 | 3 | 5 | 7 | 5 |
| 9 | 4 | 5 | 6 | 3 | 2 | 6 | 4 | 3 | 2 |
| 7 | 3 | 2 | 6 | 8 | 4 | 3 | 6 | 1 | 1 |
| 1 | 4 | 7 | 4 | 0 | 5 | 5 | 8 | 4 | 7 |
| 4 | 5 | 5 | 4 | 8 | 2 | 9 | 6 | 9 | 8 |
| 3 | 0 | 8 | 4 | 4 | 7 | 2 | 0 | 5 | 1 |
| 5 | 1 | 2 | 9 | 3 | 7 | 1 | 0 | 5 | 3 |
| 3 | 6 | 4 | 7 | 8 | 6 | 5 | 7 | 5 | 2 |
| 0 | 6 | 4 | 4 | 5 | 6 | 2 | 4 | 9 | 2 |
| 3 | 8 | 3 | 0 | 3 | 4 | 7 | 6 | 6 | 5 |

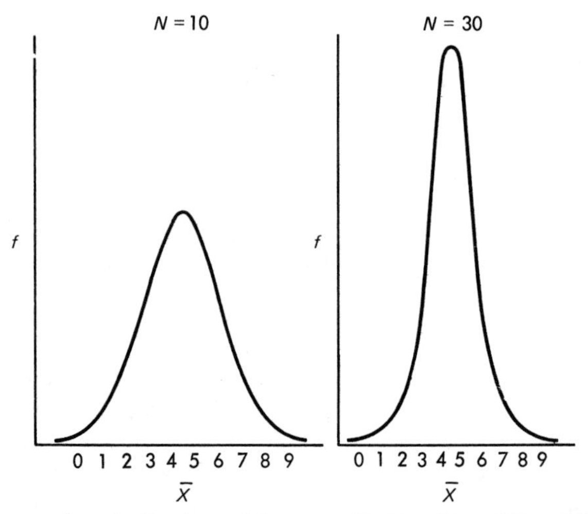


FIG. 3–3 Two sampling distributions of the mean (for N = 10 and N = 30) for the population of scores shown in Table 3-2.

The nine panels in Figure 3-5, combining our earlier examples, illustrate four basic points about the sampling distribution of the mean. (1) The variability of the sampling distribution of means is related to the amount of variability in the original population; the more variability in the population, the more variability in the sampling distribution of the mean. (2) The smaller the sample size, the more variability in the sampling distribution of the mean. (3) No matter what the shape of the original population of scores, the mean location of the sampling distribution of means approximates the value of the mean of the original population ($\mu$). (4) No matter what the shape of the original population or the size of the sample, the sampling distribution

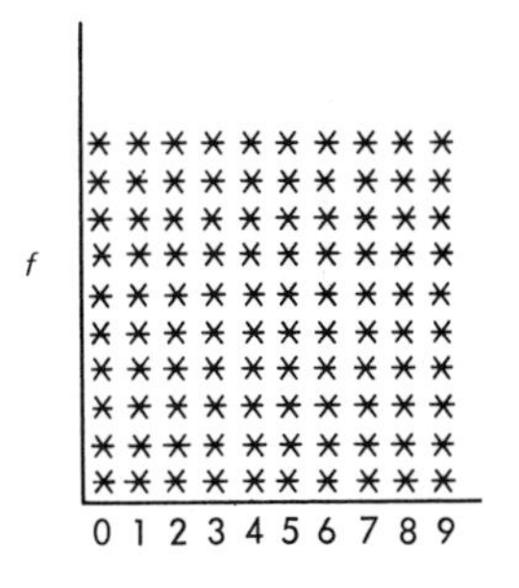


TABLE 3-3 Hypothetical population of the 100 expensive-car buyers in County C, showing the number of cars each buyer examined before making a purchase

| *Population C* | | | | | | | | | |
|---|---|---|---|---|---|---|---|---|---|
| 0 | 7 | 3 | 1 | 8 | 7 | 1 | 7 | 9 | 9 |
| 8 | 3 | 4 | 3 | 0 | 5 | 1 | 6 | 1 | 0 |
| 0 | 9 | 9 | 0 | 9 | 8 | 6 | 6 | 5 | 3 |
| 7 | 5 | 4 | 3 | 9 | 7 | 8 | 3 | 4 | 3 |
| 0 | 4 | 8 | 5 | 6 | 8 | 9 | 9 | 9 | 4 |
| 1 | 5 | 5 | 7 | 4 | 2 | 6 | 5 | 4 | 1 |
| 3 | 8 | 6 | 5 | 2 | 0 | 0 | 6 | 3 | 2 |
| 1 | 1 | 2 | 7 | 2 | 2 | 2 | 2 | 0 | 5 |
| 9 | 2 | 6 | 2 | 6 | 3 | 5 | 0 | 7 | 4 |
| 4 | 6 | 1 | 7 | 8 | 7 | 8 | 1 | 8 | 4 |

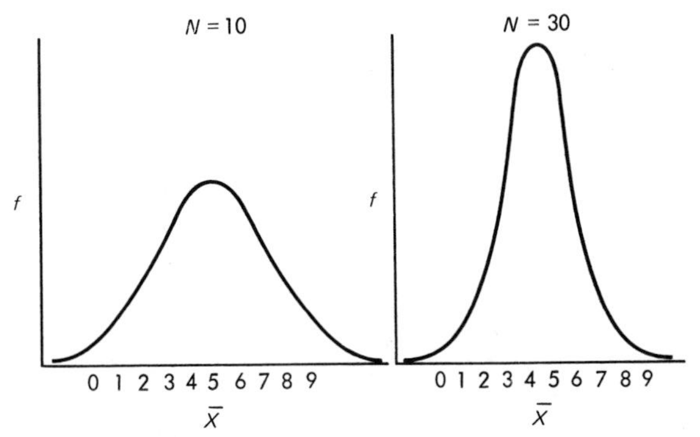


FIG. 3–4 Two sampling distributions of the mean (for N = 10 and N = 30) for the population of scores shown in Table 3-3.

of the mean tends to have the same general shape; sample means quite close to the value of $\mu$ are relatively more likely than values far away from $\mu$. Of course, it is precisely because values close to the population mean are likely that we expect a sample mean to tell us something about the population from which it was drawn.

## STANDARD ERROR OF THE MEAN ($s_{\bar{x}}$)

A sampling distribution of means is a distribution, like any other. Therefore, it has a sum of squares, variance, and standard deviation that are interrelated measures of its variability, just as any other distribution would have. The standard deviation of a distribution of sample means is often called the *standard error of the mean* ($s_{\bar{x}}$). The standard deviation of a distribution of

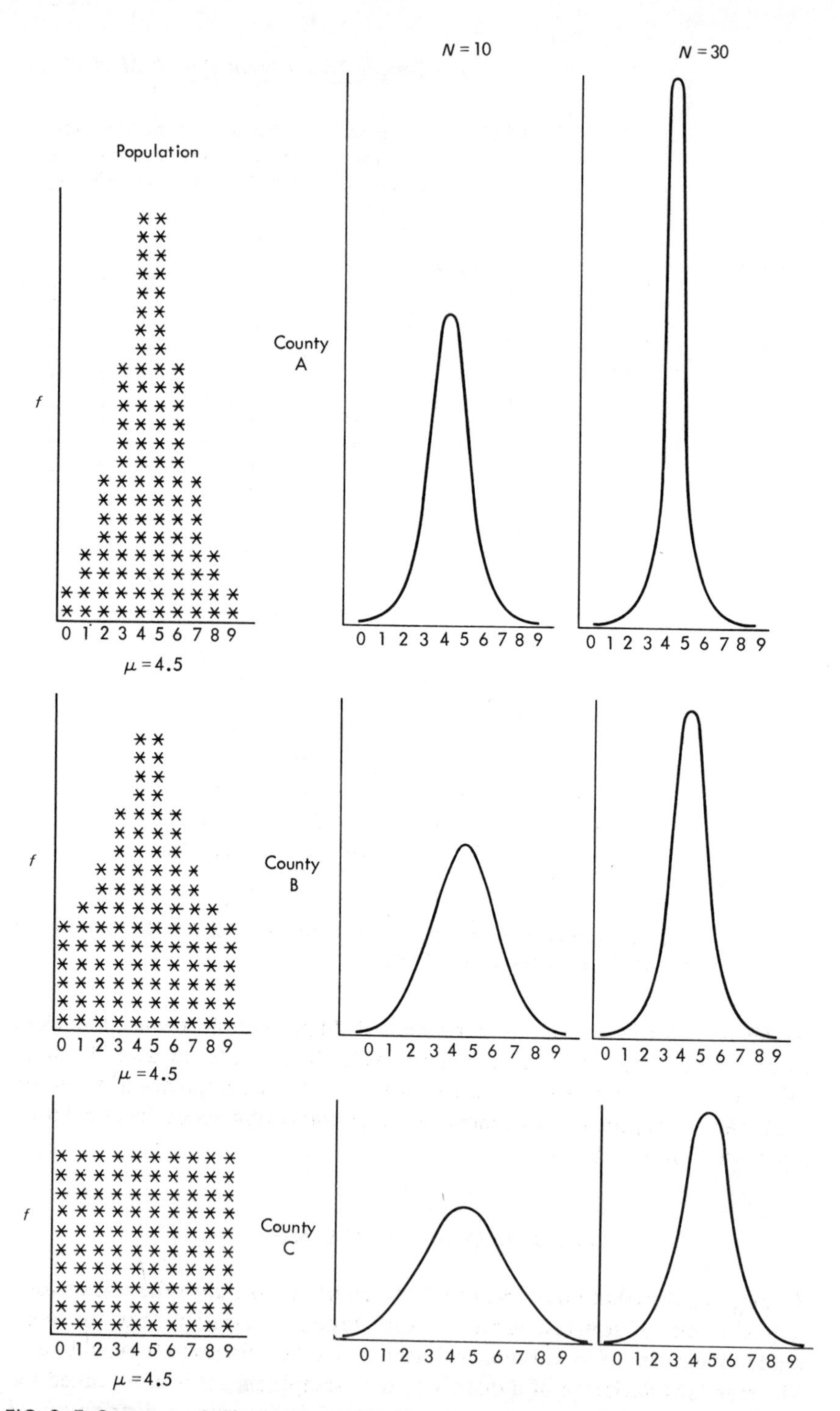


FIG. 3–5 Summary of data shown earlier for expensive-car buyers in three counties, illustrating the way in which the sampling distribution of the mean depends on the variability in the population of scores and the size of the sample.

sample means is frequently used as an index of the amount of error one might expect in using sample information as a basis for judgments about population means. We will return to this point later.

If you wanted to calculate the standard deviation of any of the sampling distributions in Figure 3-5, you might proceed in the manner outlined in Chapter 2 (pages 16–19). You would consider each sample mean as an individual observation, and you would calculate the sum of squares, the variance, and then the standard deviation. This is not, however, the most practical procedure.

### Estimating $s_{\bar{x}}$ from a Single Sample

Ordinarily, in conducting research, we would not expect to have very many sample means available from the same population. Certainly, the more samples we take, the greater the cost of any information we might obtain. However, we will often need to know the value of $s_{\bar{x}}$. Fortunately, there is a technique for estimating the standard deviation of a distribution of sample means from the data of a single sample. Remember that the variability in the sampling distribution is related to the variability in the original population ($\sigma$) and to the size of the sample ($N$). Statisticians have shown that this relationship operates according to the following formula:

$$s_{\bar{x}} = \frac{\sigma}{\sqrt{N}}$$

Look first at the numerator of this formula. It tells you that as you increase the variability in the population, you increase the variability in the sample means which might be obtained from that population. We showed this graphically in Figure 3-5. The point is illustrated again in Figure 3-6. If all the scores in the original population were the same, there would be (1) no variability in the observations and (2) no variability in the means calculated for random samples from this population. Notice that as the scores in the population become more variable, the variability of the means that might be calculated from various random samples taken from the population increases as well.

Now look at the denominator of the formula. As $N$ gets larger, $\sqrt{N}$ also gets larger, and therefore the quantity $\sigma/\sqrt{N}$ gets smaller. Thus, as $N$ is increased, the variability in the sampling distribution of the mean decreases. This, too, was shown in Figure 3-5. The point can be emphasized by thinking of the imaginary limiting case in which $N$, the sample size, equals the population size. In this case, because each sample contains the whole population, the sample mean will equal $\mu$. If the procedure is repeated, the new sample mean will also equal $\mu$, and so on. As $N$ becomes infinitely large, the variability in the sample means approaches 0. On the other hand, if $N$ is equal to 1 (in which case the mean of the sample would simply be equal to the value

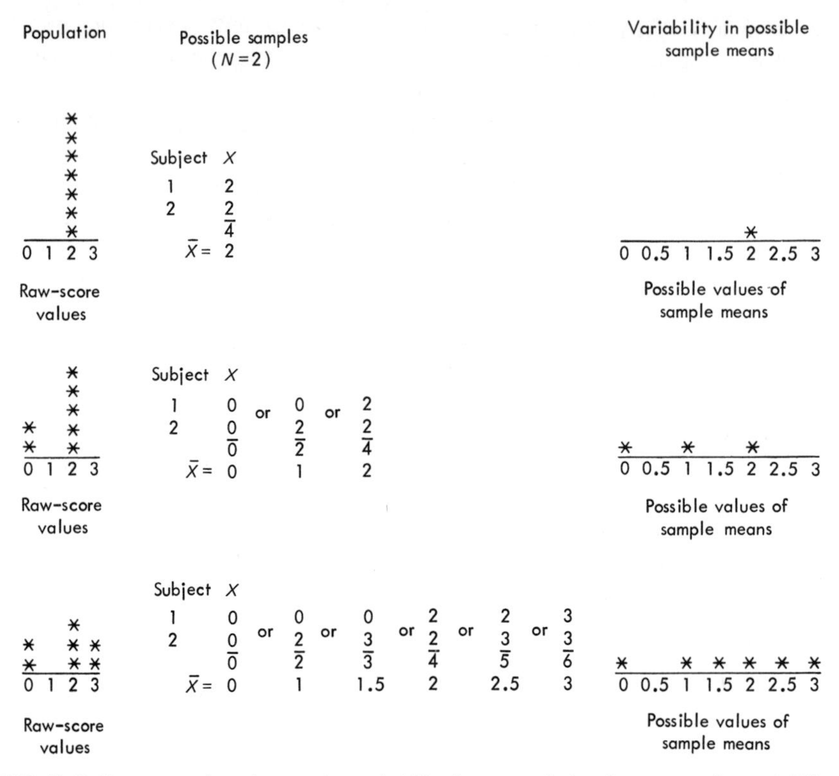


FIG. 3–6 Demonstration that as the variability in a population increases, the variability in potential sample means increases.

of the particular score drawn), the variability of the sampling distribution of the mean would be exactly equal to the variability in the original population.

Suppose we have drawn the random sample of 10 scores listed in Table 3-4 from Population A. The first thing we would do is determine the mean of this sample.

$$\bar{X} = \frac{\sum X}{N} = \frac{46}{10} = 4.60$$

Now try to think of this mean of 4.60 as being only one of many means you might have obtained, depending on which sample you happened to choose. Your confidence in the accuracy of this mean as an estimate of $\mu$, the population mean, should be related to your expectation about just how widely these possible sample means might vary. How do you assess the expected variability

TABLE 3-4 A random sample of 10 scores from Populations A and B, all the expensive-car buyers from County A and County B, respectively.

Sample from Population A ($N = 10$)

| *Subject* | $X$ | $X^2$ |
|---|---|---|
| 1 | 2 | 4 |
| 2 | 7 | 49 |
| 3 | 4 | 16 |
| 4 | 5 | 25 |
| 5 | 4 | 16 |
| 6 | 3 | 9 |
| 7 | 4 | 16 |
| 8 | 8 | 64 |
| 9 | 5 | 25 |
| 10 | 4 | 16 |
| $\Sigma$ | 46 | 240 |

Sample from Population B ($N = 10$)

| *Subject* | $X$ |
|---|---|
| 1 | 5 |
| 2 | 8 |
| 3 | 5 |
| 4 | 5 |
| 5 | 3 |
| 6 | 6 |
| 7 | 4 |
| 8 | 6 |
| 9 | 2 |
| 10 | 0 |

among sample means? First, compute the sum of squares of your sample, following the procedure shown on pages 19–20.

$$\begin{aligned} SS &= \sum X^2 - \frac{(\sum X)^2}{N} \\ &= 240 - \frac{(46)^2}{10} \\ &= 240 - \frac{2116}{10} \\ &= 240 - 211.60 \\ &= 28.40 \end{aligned}$$

Then, compute the variance of your sample, using the sum of squares and the number of observations it contains (i.e., as first shown on pages 17–19).

$$s^2 = \frac{SS}{N-1}$$

$$= \frac{28.4}{9}$$

$$= 3.16$$

You also know that

$$s = \sqrt{s^2}$$

Therefore

$$s = \sqrt{3.16}$$

$$= 1.78$$

Now, we can compute the standard error of the mean that is defined by:

$$s_{\bar{x}} = \frac{\sigma}{\sqrt{N}}$$

Your obtained $s$ is your best estimate of the variability in the original population ($\sigma$); therefore, you may substitute your obtained value into the formula for $s_{\bar{x}}$.

$$s_{\bar{x}} = \frac{\sigma}{\sqrt{N}} = \frac{s}{\sqrt{N}}$$

$$= \frac{1.78}{\sqrt{10}}$$

$$= \frac{1.78}{3.16}$$

$$= 0.56$$

Now repeat this procedure for the sample data from Population B listed in Table 3-4. Verify that $\bar{X}$ equals 4.40, $s$ equals 2.27, and $s_{\bar{x}}$ equals 0.72.

If you knew nothing whatever about these two populations other than your computed values of $\bar{X}$, $s$, and $s_{\bar{x}}$ for each sample, which sample mean would you assume is likely to involve the least sampling error? You should have greater confidence in the accuracy of the mean from Population A because its $s_{\bar{x}}$ (standard error of the mean) is smaller than that for Population B. Your reasoning should be based on the fact that a larger $s_{\bar{x}}$ indicates more variability in potential means, implying that values of the sample mean which deviate substantially from the population mean are more likely to occur.

In addition to the amount of variability in the original population, another factor, sample size, influences our confidence in the accuracy of any particular sample mean. Converting means to standard scores helps make this clearer. Just as a raw score can be expressed as a standard score, a mean can be expressed as a standard score. The raw score is more meaningful in relation to the mean and standard deviation of the distribution from which it comes. Similarly, a mean is more meaningful in relation to a distribution of other means that might be obtained under exactly the same circumstances.

The procedure for expressing a sample mean in terms of its deviation from the most typical score in a distribution of means is exactly analogous to that described for raw scores in Chapter 2. For example, for the car buyers in County A, what would be the standard score of a mean of 6 in a distribution of means from samples of size $N = 10$?

Recall that the mean of Population A is equal to 4.50; therefore, the mean of the sampling distribution in Figure 3-1 should also be 4.50. We have already estimated the standard deviation of this sampling distribution, and it is 0.56. Therefore, when $N = 10$

$$\begin{aligned}\text{Standard score for mean of 6} &= \frac{6.00 - 4.50}{0.56} \\ &= \frac{1.50}{0.56} \\ &= +2.68\end{aligned}$$

What would be the standard score of a mean of 6 from Population A when $N = 30$? This problem requires that we know the standard deviation of the sampling distribution of means when $N = 30$ (see Figure 3-2). Assume that your computed value of $s$ equals 1.78 for a random sample of 30 scores. Then

$$\begin{aligned}s_{\bar{x}} &= \frac{s}{\sqrt{N}} \\ &= \frac{1.78}{\sqrt{30}} \\ &= \frac{1.78}{5.48} \\ &= 0.32\end{aligned}$$

Now, compute the standard score for a mean of 6 when $N = 30$.

$$\text{Standard score} = \frac{6.00 - 4.50}{0.32}$$

$$= \frac{1.50}{0.32}$$

$$= +4.69$$

By comparing these two standard scores (+2.68 and +4.69), it is obvious that a sample mean of 6 from Population A would be a less frequent occurrence with a sample of 30 cases than it would with a sample of 10 cases. We will return to this point in Chapter 4.

## Practice Problems

### A. *What Are They Worth?*

The U.S. Senate decided to commission the Gullible Poll Company to determine what salary a senator ought to be paid. Gullible randomly sampled 100 adults and reported the mean salary suggested. The Senate was not happy with the outcome and paid Gullible to take another random sample of 100 adults. This time, the mean salary was higher, and so the senators were sure that the second sample more accurately reflected the views of the public.

1. Which was the better sample?
2. If Gullible repeated this procedure a very large number of times, what distribution would be obtained?

### B. *Soybeans, Anyone?*

The Grow-Good Soybean Company hires you to collect information about potential markets for their protein products. Suppose the mean number of pounds of meat eaten each year by an adult in each of two countries (Northmunch and Globia) is as given below (the standard deviations are also listed):

| Northmunch | Globia |
|---|---|
| $\mu = 60, \sigma = 12$ | $\mu = 60, \sigma = 24$ |

1. Describe the sampling distribution of the mean for samples of size $N = 144$ for these two countries as fully as you can.
2. What is one way to decrease the variance of the sampling distribution of the mean from Globia? How large a sample would you need in order to have $s_{\bar{x}} = 1.00$?
3. On a trial basis, you provide free soybeans to four randomly selected households in Globia. The scores that follow are the number of soybean main dishes served in each household during the last month of a six-month period. Estimate the standard error of the mean for these data.

| $X$ |
|---|
| 17 |
| 20 |
| 23 |
| 16 |

## *C. The Numbers Game*

The variability of a sampling distribution of means is generally (less than, equal to, or greater than) the variability of the scores from the original population of raw scores. Why?

# Theoretical Distributions and Probability

# 4

SUPPOSE THAT 1,000 PEOPLE WERE SELECTED RANDOMLY from the total population of adults in the United States, that the weight of each was determined, and that their weights were entered on a graph to indicate the relative frequency of each weight. The result would look very much like the distribution shown in Figure 4-1. Notice the shape, highlighted by the smooth, heavy curve of its outline drawn along the top. From the curve, we see that most of the scores are near the center, with fewer and fewer scores as you move farther away in both directions.

The curve in Figure 4-1, often called the *normal curve*, is important because it also describes a very large number of naturally occurring distributions. These include such diverse phenomena as the frequency of errors on most school examinations, scores on most intelligence tests, and average temperatures on January 1 for any city in the world, to name just a few.

The mathematical equation that describes normal curves was probably first derived in 1733 by a mathematician, Abraham Demoivre, who helped wealthy London gamblers solve problems of chance. Later, in 1786, the equation was discovered independently by the French astronomer and mathematician Pierre Laplace; and in 1809, a German astronomer and mathemati-

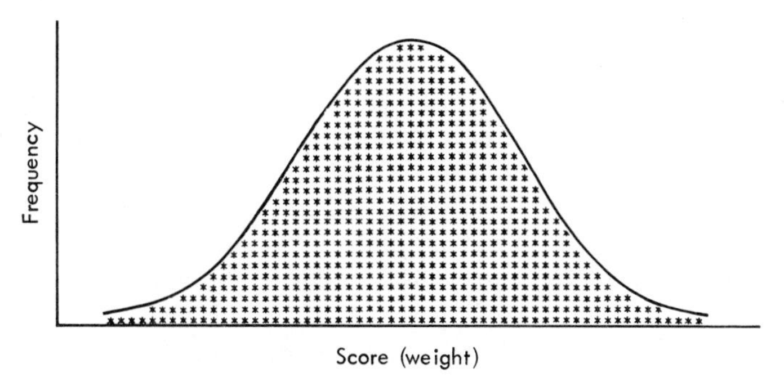


FIG. 4–1 Hypothetical distribution of the weights of 1,000 randomly selected adults in the United States, illustrating a normal distribution.

cian, Karl Friedrich Gauss, offered a new derivation of the formula. For these astronomers, the equation was interesting as a model of errors made in astronomical observations and scientific observations in general. In 1835, a mathematician and astronomer from Brussels, Lambert Quételet, suggested that this set of curves was widely applicable to the description of human physiological and social traits. He found, for example, that a distribution of the heights of French soldiers looked very much like the distribution shown in Figure 4-1, and he believed that the distributions of many other traits would also show this same general shape when the data were collected. Quételet proposed that the "average" man was the ideal of nature and that deviations from the average represented nature's mistakes, with small mistakes being more frequent than large ones. This is very similar to Gauss's ideas about the mistakes astronomers make in their scientific observations. The equation for normal curves is commonly referred to as the *normal law*.

## NORMAL CURVES

Often, you will hear references to *the* normal curve. Actually, the equation for normal curves defines a whole family of curves.

$$f(X) = \frac{1}{\sigma\sqrt{2\pi}} e^{-(X-\mu)^2/2\sigma^2}$$

Notice that this equation contains several constants, including $\pi$ and $e$.* There are two components of this equation that may vary from curve to

*Recall that $\pi = 3.14$ and $e = 2.72$.

curve but that are fixed in value for the distribution of any particular population. These are the distribution mean ($\mu$) and the standard deviation of the distribution ($\sigma$). In other words, there is a different normal curve for each different pair of values of $\mu$ and $\sigma$. Thus, there is not one normal curve; rather, there are an infinite number of normal curves, one for each potential combination of a particular mean and standard deviation. Some examples of normal curves are given in Figure 4-2. In Figure 4-2a, you can see three normal curves with the same mean and different standard deviations; in Figure 4-2b, three curves with the same standard deviations and different means.

The similarity in shape of all these curves depends on the specific relationship of the components in the normal-curve equation. Furthermore, normal

(a) Normal curves with the same mean and different standard deviations.

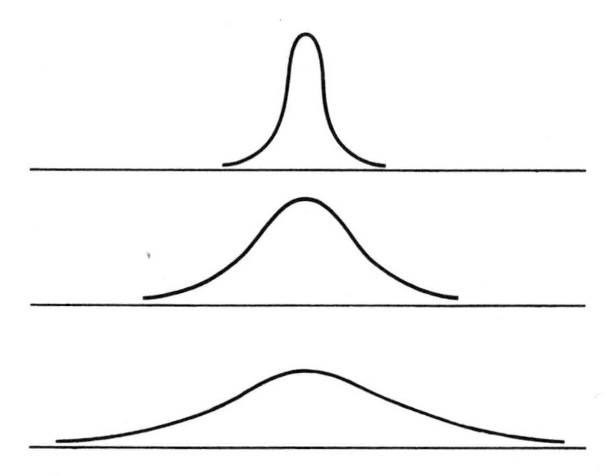

(b) Normal curves with the same standard deviation and different means.

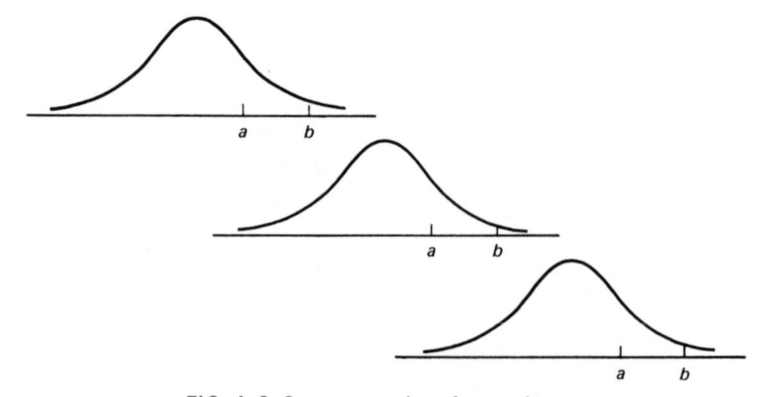


FIG. 4–2 Some examples of normal curves.

curves, no matter what the specific values of their means and standard deviations, have certain properties in common. Some of these properties are more obvious than others. For example, in any particular normal curve, the mean, median, and mode have the same value. All normal curves are symmetrical and have a bell-like shape.

For our purposes, a more important (but less obvious) property common to all normal curves is related to the area under the curve. The area under any curve represents 100 percent of the cases in the distribution, and the area under a curve between any two points on the horizontal axis (e.g., $a$ and $b$ in Figure 4-2b) is the percentage of the cases in the distribution with scores between the values at those points. By using the equation for a curve and some knowledge of calculus, you can determine the area between any two points. If these points are expressed in standard-deviation units rather than raw scores, then the area between any two of these points is always the same for all *normal* curves and can be obtained very easily from a table. Now we will demonstrate how this is done.

Look first at Figure 4-3a. This is a distribution with a mean of 20 and a standard deviation of 4. Compute the standard score that corresponds to a raw score of 24 in this distribution.

$$\text{Standard score} = \frac{X - \bar{X}}{s} = \frac{24 - 20}{4} = +1.00$$

Now compute the standard score for a raw score of 1 in the distribution in Figure 4-3b.

$$\text{Standard score} = \frac{1 - 0}{1} = +1.00$$

Verify that a raw score of 42 in Figure 4-3c has a standard score of $+1.00$ and that a raw score of 64.5 in Figure 4-3d has a standard score of $+1.00$.

Now turn to Appendix C (page 217) for a table of areas under normal curves for different points. If we let the computed value of the standard score in a normal distribution equal $z$, in each of the cases above, $z = +1.00$. Look down the column of the table headed $z$ until you find 1.000. The corresponding number in the table is 0.3413 (or 34.13 percent). This is the proportion (or percentage) of total area under a normal curve that is between the mean and a standard score of $+1.00$. This value is designated in each of the four curves in Figure 4-4.

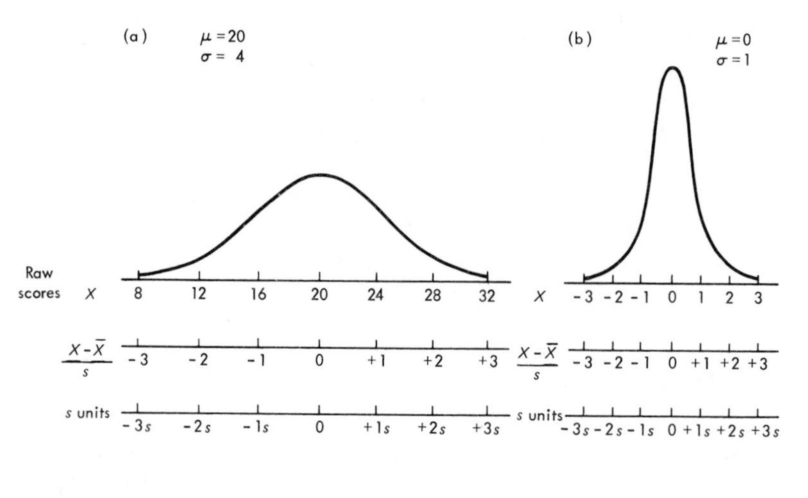


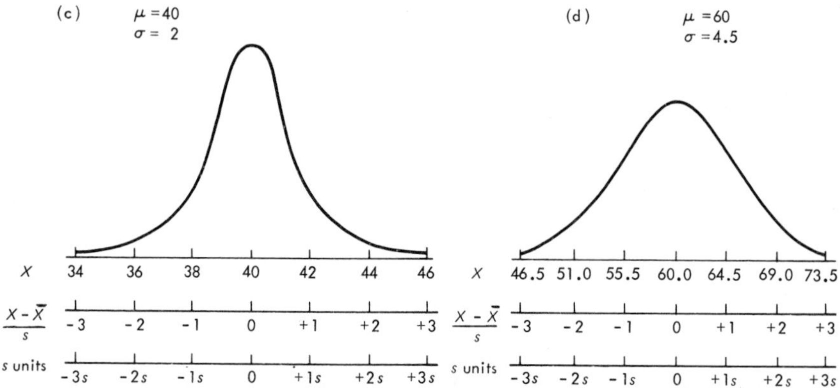


FIG. 4–3 Four normal curves and corresponding standard scores for each.

Now look at the area between the mean and a standard score of +2.00 in each distribution. According to the table in Appendix C, this area is 0.4772 of the total. That is, in any normal curve, 47.72 percent of the scores fall between the mean and 2.00 standard-deviation units above the mean. This value is also indicated for each curve in Figure 4-4.

Suppose you wanted to know the proportion of the area of the curve between a standard score of +1.00 and a standard score of +2.00. This area could easily be found by subtraction.

| | |
|---|---|
| 0.4772 | area between mean and +2.00 standard-deviation units |
| −0.3413 | area between mean and +1.00 standard-deviation units |
| 0.1359 | area between +1.00s and +2.00 standard-deviation units |

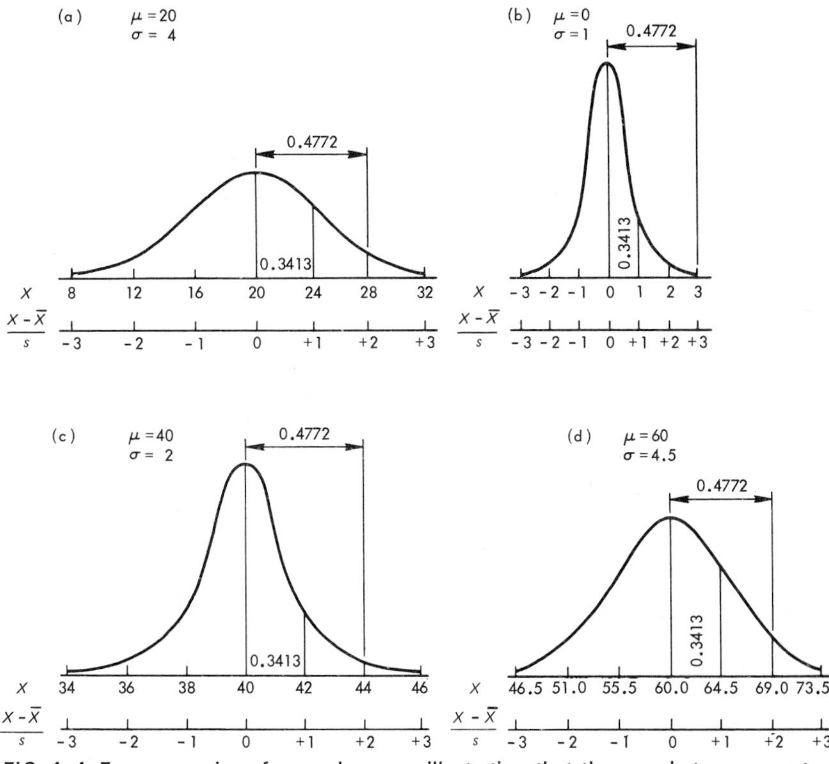


FIG. 4–4 Four examples of normal curves, illustrating that the area between any two particular standard-score values is the same for any normal curve.

This value, 0.1359, appears in the appropriate place on the curves in Figure 4-5. Now, using the same procedure, verify that the area between the mean and a standard score of +3.00 is equal to 0.4986 and that the area between standard scores of +2.00 and +3.00 is 0.0214. Because the mean of any normal curve cuts the curve in half, 50 percent of all the scores occur above the mean. As you can see, almost all of these (49.86 percent) are between the mean and +3.00 standard-deviation units. Only 0.14 percent are above this point.

Appendix C lists only the areas for positive standard scores. But because normal curves are symmetrical, areas below the mean (for standard scores with negative signs) correspond exactly to their counterparts above the mean. The appropriate areas have been indicated for points below the mean on the curves in Figure 4-5.

Figure 4-5 also illustrates that the area between the mean and any particular *standard score* is the same in *all* normal distributions. Because this is

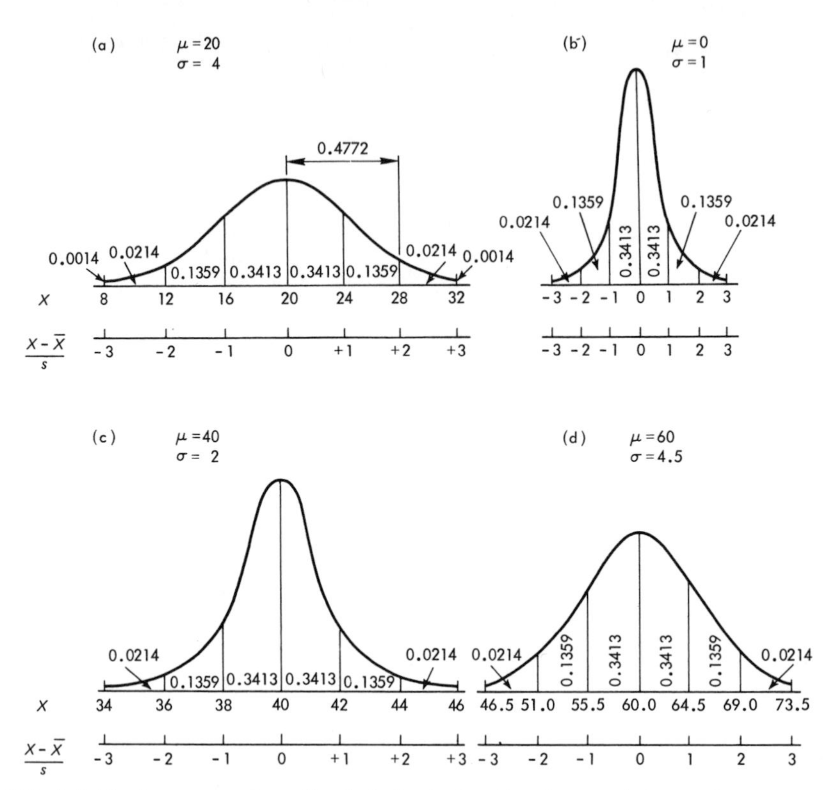


FIG. 4–5 The four curves from Fig. 4–4, illustrating that the area between the mean and any particular standard score is the same regardless of whether the standard score is negative or positive.

the case, the table in Appendix C (which was obtained with the help of calculus) can be used in determining areas under all normal curves.

### Applications of Normal Curves

*Determining the probability of values of raw scores.* Suppose you want to try out a program that you think will improve reading comprehension of fifth-grade students. The program is time consuming and requires you to work with each child individually; therefore, the number of youngsters you can admit is limited. You want to try the program out on children who are relatively poor readers because they have the most to gain from any program that works. On the other hand, you know from past experience that some children have extremely severe reading problems and that they are unlikely to show any improvement over the time interval covered by your program.

You also have a test to assess the current level of reading ability. By using this test, you determine that the reading scores of a population of 5,000 fifth-grade students in a certain state are normally distributed, with $\mu = 121$ and $\sigma = 20$. You decide to choose subjects on the basis of these pretest reading scores so that you can select a fairly homogeneous sample of children with scores between 110 and 115. You know that if you restrict the range of abilities from which you sample, this will limit the generality of your results to only those students with pretest scores from 110 to 115. But at this point, you do not mind this limitation because your major goal is to determine whether your program can help some children.

Because the children's pretest scores are normally distributed, we can use the table in Appendix C to answer the following questions.

1. What is the probability of selecting a child *at random* from the population of 5,000 who has a reading score between 110 and 115?

This problem is diagrammed in Figure 4-6. You need to determine the

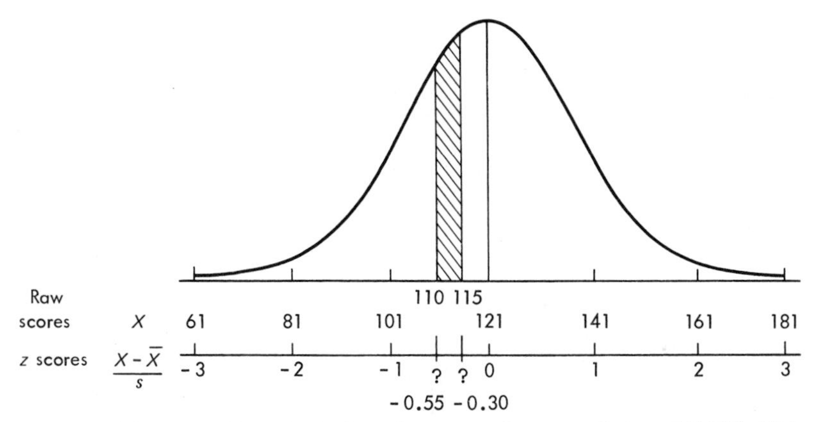


FIG. 4–6 Distribution of hypothetical reading scores for a population of 5,000 children. The problem is to determine the area under the curve between scores of 110 and 115. This requires calculating the standard scores represented by the question marks.

area of the curve between the scores of 110 and 115. The corresponding standard scores are computed below.

$$z = \frac{110 - 121}{20} = \frac{-11}{20} = -0.55 \qquad z = \frac{115 - 121}{20} = \frac{-6}{20} = -0.30$$

The relevant area can now be determined using the table in Appendix C. The negative $z$ scores are treated as if they are positive.

| | |
|---|---|
| 0.2088 | area between mean and −0.55 standard-deviation units |
| −0.1179 | area between mean and −0.30 standard-deviation units |
| 0.0909 | area between −0.55 and −0.30 |

Thus, the probability of choosing a child with a score between 110 and 115 is 0.0909.

2. What percentage of the children in this population would qualify to be in your program, given the way you have defined your population of interest?

$$0.0909 \times 100 = 9.09 \text{ percent}$$

3. How large is your population of interest? The answer can be found by multiplying the total population (5,000) by the proportion who would qualify (.0909). Therefore

$$5{,}000 \times .0909 = 454.50$$

You should not be misled by the fact that we have devoted a good deal of space to normal curves. In spite of Quételet's faith, and in spite of the large number of characteristics that were found to be normally distributed once the collection of large amounts of data became more common, not every aspect of the world is normally distributed. For example, income among the population of adults in the United States probably looks more like the curve in Figure 4-7. But even though not all characteristics are normally distributed, many statistical indices are normally distributed under a wide range of conditions, and this is an extremely useful fact.

*Determining the probability of specified values of the sample mean* ($\bar{X}$). In Chapter 3, you saw several examples of the sampling distribution of means. Recall that this is the distribution of values of $\bar{X}$ which would be obtained with repeated random samples of size $N$ from a particular population of scores. In Figure 3-5, you saw that the sampling distribution of the mean tended to be symmetrical and bell-shaped. In fact, when $N$ is large enough, the sampling distribution of the mean will be very close to a normal distribution. This is true even when the original population from which the scores are sampled has a nonnormal distribution. The fact that under a wide range of conditions the sampling distribution of the mean is normally distributed is very useful in practical research. Right now, we can use this fact to make our previous discussion of the sampling distribution of the mean more precise.

Suppose that you randomly sampled 36 subjects from the population of 5,000 fifth graders described in the last example and recorded their pretest scores. What is the probability that you would obtain a sample mean either above 126 or below 116?

This problem is diagrammed in Figure 4-8. The solution involves determining the shaded area of the curve for the sampling distribution of the mean for $N = 36$. Because the sampling distribution of the mean is approximately

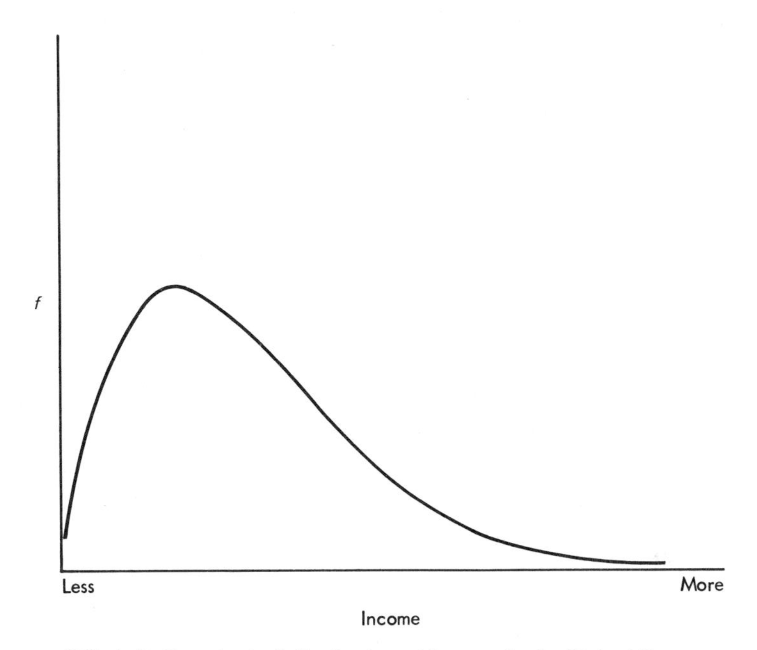


FIG. 4–7 Hypothetical distribution of income in the United States.

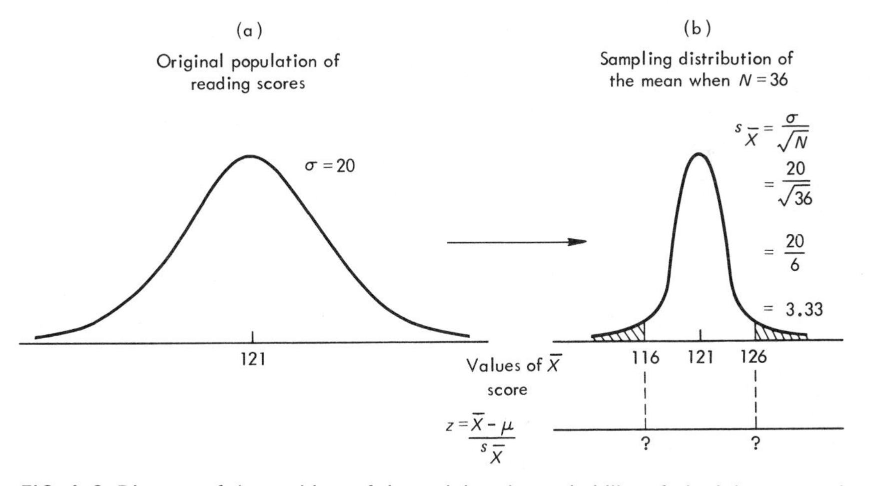


FIG. 4–8 Diagram of the problem of determining the probability of obtaining a sample mean above 126 or below 116 when $\mu$ = 121, $\sigma$, 20, and sample $N$ = 36.

normally distributed when $N = 36$, we can use the normal model to obtain the answer.

First, we need to compute standard scores corresponding to the raw means of 126 and 116. We also know that the mean of a sampling distribution of means should be equal to the mean of the original population of scores (which was 121) and that the standard deviation of a sampling distribution of means should be equal to $\sigma/\sqrt{N}$. In this case, $\sigma = 20$ and $\sqrt{N} = \sqrt{36} = 6$; $\sigma/\sqrt{N} = 3.33$. Therefore, the necessary standard score for $\bar{X} = 126$ is

$$\frac{126 - 121}{3.33} = +1.50$$

and the necessary standard score for $\bar{X} = 116$ is

$$\frac{116 - 121}{3.33} = -1.50$$

According to the table in Appendix C, the area between the mean and a standard score of 1.50 is 0.4332. Because exactly half, 0.5000, of the area is above the mean (remember, normal curves are symmetrical), the area above 126 can be obtained by subtraction.

$$\begin{array}{r} 0.5000 \\ -0.4332 \\ \hline 0.0668 \end{array}$$

The area below a $z$ score of $-1.50$ would be the same as the area above $+1.50$. Therefore, the probability of obtaining a mean greater than 126 *or* less than 116 would be $0.0668 + 0.0668$, or 0.1336. Another way of saying the same thing is that in approximately 13 samples out of 100 (13 percent), you would expect to obtain a sample mean as deviant from the true mean (121) of the population as 126 or 116.

What would be the probability of obtaining a sample mean as large as 126 or as small as 116 if the sample size were 100 instead of 36? Verify that the answer is 0.0124. Be sure to diagram the problem as shown in Figure 4-8.

In Chapter 3 (Figure 3-5), you saw that increasing the sample size decreases the variability of the sampling distribution of the mean. The present problem emphasized that one very important consequence of increasing the size of a sample is that the probability of extremely deviant values of the sample mean decreases. Thus, as sample size increases, our sample mean is more likely to be close to the true value of the mean. This is clearly more important when we do not know the true value of the population mean (the state we are in in most actual research) and must rely on the information in our sample for our opinions and conclusions.

## *t* DISTRIBUTIONS

In the preceding section, we said that the normal curve is particularly useful to researchers because many statistical indices, such as standard scores for a sampling distribution of means, are normally distributed. In fact, this statement holds only under some circumstances. For example, the normal-curve equation closely approximates the sampling distribution of the mean only when the samples are reasonably large, say, $N = 30$ or more. However, when samples are smaller than this, and when $\sigma$ must be estimated from our sample data, the sampling distribution of the mean tends to have fewer observations clustered near the mean of the population and more cases at the extremes of the distribution than you would expect on the basis of a normal-curve model.

The fact that the sampling distribution of the mean is not distributed normally for small $N$ was first pointed out in 1908 by a British statistician named Gosset. He derived an equation that specifies a set of distributions which more closely fit the sampling distribution of the mean with small samples drawn from populations whose standard deviations must be estimated. He called these distributions *t distributions*, and because he published under the pseudonym of "Student," these distributions and the statistical test associated with them are sometimes called *Student's t*.

Like a normal distribution, any particular $t$ distribution depends upon $\mu$ and $\sigma$. But unlike the family of normal curves, a particular $t$ distribution also depends upon the number of observations in each sample. There is a different $t$ distribution for each sample size; for example, there is one $t$ distribution that correctly characterizes the distribution of sample means drawn from a population in which each sample consists of 10 observations and a different $t$ distribution that correctly characterizes the distribution of means from samples of 20 observations. The practical consequence of this is that the area between the mean of the sampling distribution and any particular sample mean (expressed as a standard score) is not always the same; rather, it depends on the sample size.

Figure 4-9 indicates what would happen if you used a normal curve as a

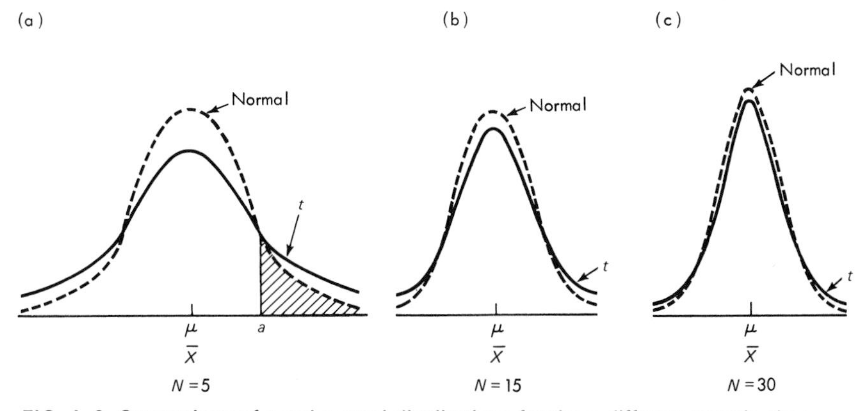


FIG. 4–9 Comparison of $t$ and normal distributions for three different sample sizes.

basis for estimates of the relative probabilities of various values of a sample mean for samples in which $N = 5$, 15, or 30. In Figure 4-9a, point *a* corresponds to the standard score computed for a particular value of the sample mean. The striped area under the broken curve indicates the area above *a* (obtained with the help of the normal-curve table in Appendix C). The solid curve indicates the actual shape of the sampling distribution of the mean when $N = 5$. The shaded area indicates that positive standard scores at least as extreme as *a* are somewhat more probable than indicated by the normal-curve values. Figure 4-9b and 4-9c indicate that as sample size increases, the extent to which a normal distribution underestimates the variability of the standard scores of sample means decreases. As the sample size gets larger, the normal distribution becomes a more accurate model. The usual rule of thumb is that when $N = 30$ or more, the error involved in assuming that the sampling distribution of the mean is a normal distribution rather than a *t* distribution is trivial.

Because the area between the mean of the sampling distribution and the particular sample mean is not always the same, there is no single table for *t*, as there is for normally distributed scores. Rather, there is a different table for each possible sample size. To conserve space, only a few important (frequently needed) areas of each *t* distribution are included in Appendix D. The information in this table is very similar to that provided in Appendix C, but the table is set up a little differently.

To use the table, you must first locate the appropriate distribution according to the sample size. Actually, the correct distribution is usually specified, not by *N* exactly, but by what is called the *degrees of freedom* (*df*) for a particular sample. We will not explain the concept of degrees of freedom in detail, but you should know that *df* is always directly related to sample size. In the case of the sampling distribution of the mean, the degrees of freedom is $N - 1$. You should note the fact that $N - 1$ is also the value of the denominator in computing $s^2$ ($s^2 = SS/N - 1$).

In order to illustrate the use of Appendix D, we will describe a concrete example. Suppose you sailed to the Island of Lith, randomly sampled 9 female sea turtles, counted the eggs in the nest of each subject, and obtained $\bar{X} = 113.8$ and $s = 18$. How likely would you be to obtain a mean at least this large if the true mean of the population were 100? The problem is diagrammed in Figure 4-10.

First, assume that in the original population of scores (number of eggs in each nest), $\mu = 100$ and $\sigma = 18$. What would the sampling distribution of the means from successive random samples of $N = 9$ look like? First, it should have a mean of 100 because the central tendency of the sampling distribution of the mean is the true mean of the population. Second, it should have a standard deviation $(s_{\bar{x}}) = \sigma/\sqrt{N} = 18/\sqrt{9} = 6$. Next, the standard

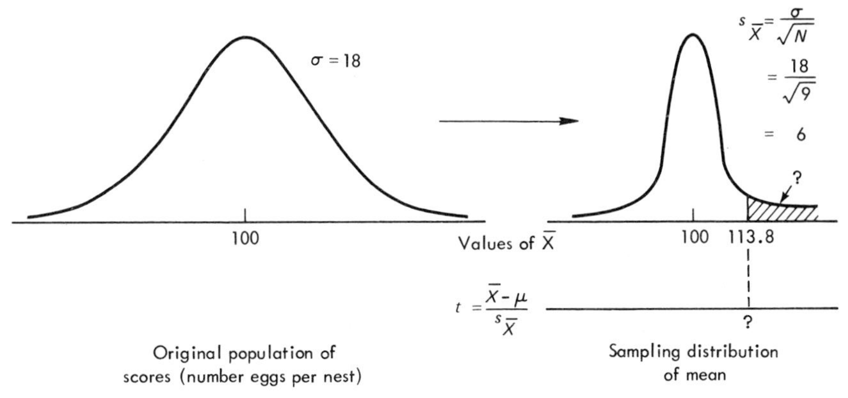


FIG. 4–10 Diagram illustrating how to determine the likelihood of obtaining a sample mean of 113.8 or more when the population mean is actually 100 and $N = 9$.

score corresponding to an obtained mean of 113.8 is computed as follows:

$$\text{Standard score} = \frac{113.8 - 100}{6}$$
$$= \frac{13.8}{6}$$
$$= 2.30$$

Now, what you want to determine is the area above a standard score of +2.30. Because the standard scores of small samples fall into a $t$ distribution rather than a normal curve, the area between the mean of the distribution and a standard score of 2.30 will depend on the size of the sample and, more particularly, on the degrees of freedom associated with the sample. In this case, the sampling distribution of the mean is distributed as $t$ with

$$df = N - 1 = 8.$$

Now turn to Appendix D. Look down the left-hand column until you have found the row for $df = 8$. This row comprises the information you have about the $t$ distribution with $df = 8$. If you look at the column labeled $t_{.10}$, you will see an entry of 1.397. This indicates that .10 (or 10 percent) of the area under the curve is to the right of a standard score of 1.397. The entry in the next column indicates that .05 (5 percent) of the area is to the right of a standard score of 1.860. The entry in the next column indicates that .025 of the curve is above a standard score of 2.306, and so on. Notice that our obtained standard score of 2.30 is just a shade closer to the mean than the standard score (2.306) above which 2.5 percent of the standard scores should fall. Therefore, 2.5 percent is a very close estimate of the percentage of sample means that should be as large as or larger than our obtained sample mean if the true mean of the population equaled 100. To obtain a closer estimate of the area above 2.306, we would need a more detailed table (much like the normal-curve table) for each $t$ distribution. In practical applications, we rarely need more detail than that contained in Appendix D.

Suppose you had assumed that the sampling distribution of the mean was normally distributed when $N = 9$. What would have been your estimate of the probability of obtaining a mean at least as large as 113.8? According to Appendix C, the area between the mean and a standard score of 2.30 is $0.5000 - 0.4893 = 0.0107$. Your estimate of the chances of obtaining a mean greater than 113.8 from a distribution whose true mean was 100 would have been approximately 1 in 100 if you had assumed the sampling distribution was normal. Assuming that the distribution is more like a $t$ distribution with 8 *df* indicates that the chances are closer to 2.5 in 100. Thus, if you had used a normal-curve model in this case, you would have assumed that your obtained mean was more unusual than it really was. Of course, as the sample size increases, estimates based on the appropriate $t$ and a normal model will be increasingly similar.

These model distributions are not merely of theoretical or mathematical interest, of course. They form the basis for many of the practical statistical tests that we will discuss in the remaining chapters of this book. An understanding of them is quite important to seeing how statistics works as a tool in research.

## Practice Problems

### *A. Running Fast*

All the stopwatches of a given model from Company A's inventory were compared with an accurate standard watch. Generally, they tended to run fast. The number of seconds each watch exceeded the one-minute test interval was recorded. The mean of the resulting distribution was 7.30 seconds, and the variance was 4.00.

1. Assuming the error scores are normally distributed, what percentage of the watches are fast by 5 seconds or more?
2. Company B has a similar model for the same price. These watches were measured by the same procedure used for Company A's watches, and they also tended to run fast. The distribution was normal, with a mean of 10.00 seconds and a variance of 0.01. Which company would you rather have supply stopwatches for your laboratory? Why?

## *B. How "Qualified"?*

A large company is recruiting employees from colleges in Michigan. All applicants are required to take a test that assesses general management skills. The company has a policy of rejecting applicants with scores of 95 or less (as "underqualified") and also has a policy of rejecting applicants with scores of 103 or more (as "overqualified"). The number of students taking the test this year was 500. The scores were normally distributed with a mean of 100.50 and a variance of 16.00.

1. What *percent* of the applicants were underqualified?
2. What *percent* of the applicants were overqualified?
3. How *many* acceptable applicants were there?

## *C. Protein Again*

Review the information in Practice Problem 3-B (page 38). If $N = 144$, what should be the probability of obtaining a sample $\bar{X}$ less than 58 or greater than 62 from the Northmunch population? From the Globia population? How can these two probabilities be different when the true means of the populations are the same?

## *D. The Numbers Game*

See Practice Problem 3-B again. If the true mean number of soybean dishes served in a month for the total population of Globia households was $\mu = 20$, what values of $\bar{X}$ for $N = 4$ would include all but the 10 percent most deviant possible sample means? *Hint:* First decide what theoretical distribution your sampling distribution will resemble with $N = 4$.

# Comparing Two Populations: Principles and Computation

# 5

SUPPOSE THAT AN EDUCATIONAL PSYCHOLOGIST was asked whether small classes are more effective than large classes in teaching algebra to high school students. Certainly, one might guess that learning is more efficient when classes are small, inasmuch as a teacher can give more personalized attention to each student's needs when the overall demands on his or her time are reduced. But guesses, even quite "educated" guesses, are no substitute for direct evidence, and so the psychologist might choose to perform an experiment that would provide a more certain answer. How might such an experiment be set up?

As is the case with most inquiries, a first basic step is to state the problem as clearly as possible. In research, often the best way to determine the consequences of some proposed change in the way things are done (e.g., reducing the size of high school algebra classes) is by stating the effect we think the change will produce and then seeing whether our expectation actually holds up under test conditions. The investigator thus advances a testable *experimental hypothesis* that concisely summarizes the proposition in which he or she is interested. In our example, the basic experimental hypothesis might simply state: "Small high school algebra classes will lead to more complete learning than large ones."

Once an experimental hypothesis is formulated, the next step is to subject it to direct test in a *controlled situation.* A controlled situation, which the experimenter creates, must include each of the components required to test the basic hypothesis and must eliminate all factors that are not relevant. With respect to our experimental hypothesis, then, the experimenter must create two types of algebra classes, large and small, and have some students enroll in each. It will be necessary to decide what is meant by a "large" class and by a "small" one, so let us suppose, somewhat arbitrarily, that a large class is one containing 30 students and a small class is one containing 10 students.*

Having defined our problem more clearly, it is then necessary to take a sample of prospective algebra students and divide them into two groups, those who will be taught in a class of 30 students and those who will be taught in a class of 10 students. Assignment to these two treatment groups would be made randomly for the reasons discussed in Chapter 2.

Finally, we must find a way to measure the variable of interest to us, amount of algebra learned, so that it can be related to class size. A standard examination for all students, to be given at the end of the semester, seems well-suited to the task and will yield scores that can be statistically analyzed in testing our hypothesis.

But how, more precisely, would we use statistics in this case? We can begin by examining the obtained scores and their means, as shown in Table 5-1. Notice that, as expected from the hypothesis, *on the average* those in the 10-student classroom learned more than those in the 30-student classroom. However, even a very large mean difference occurring in one comparison might not reflect a "real" advantage for smaller classes. Our suggestion that a much larger mean score for the 10-student sample does not automatically demonstrate an advantage in small classes may seem surprising at first, but it does make sense when several additional facts are considered.

The first point is that the average difference between two groups receiving different treatments will almost never be 0, *even if the treatment difference had no effect whatsoever.* (Recall that we have already seen, in Chapter 3, that two samples drawn randomly from the same population will almost always differ to some extent; the point being made here is analogous.)

*The problem of translating broad ideas and imprecise labels into concrete terms is a thorny one. Who is to say, for example, that a class of 14 students does not fill the definition of "small" or that a class of 25 is not quite "large"? In practice, an investigator who is initially testing a hypothesis will select a very clear contrast in treatment between groups in order to see if any difference at all is obtained; subsequently, if initial results warrant it, less glaring changes might be tried to determine the limits of the hypothesis. We shall have more to say about such decisions later.

TABLE 5-1 Hypothetical scores on algebra test of students randomly assigned to a small or a large class

| *Subject* | *Small (10-Student Class)* | *Large (30-Student Class)* |
|---|---|---|
| 1 | 87 | 98 |
| 2 | 100 | 87 |
| 3 | 89 | 70 |
| 4 | 95 | 94 |
| 5 | 63 | 80 |
| 6 | 81 | 76 |
| 7 | 84 | 79 |
| 8 | 79 | 64 |
| 9 | 78 | 46 |
| 10 | 34 | 40 |
| 11 | 92 | 88 |
| 12 | 90 | 90 |
| 13 | 48 | 78 |
| 14 | 86 | 80 |
| 15 | 75 | 65 |
| 16 | 96 | 43 |
| 17 | 89 | 76 |
| 18 | 86 | 83 |
| 19 | 73 | 72 |
| 20 | 77 | 81 |
| 21 | 56 | 90 |
| 22 | 84 | 85 |
| 23 | 93 | 100 |
| 24 | 87 | 74 |
| 25 | 72 | 68 |
| 26 | 86 | 25 |
| 27 | 98 | 47 |
| 28 | 88 | 71 |
| 29 | 70 | 80 |
| 30 | 64 | 86 |
| $\sum X$ | 2,400 | 2,216 |
| $\bar{X}$ | 80.00 | 73.87 |

The second point to notice is that the experimenter is usually interested, not in sample differences per se, but rather in whether sample differences, when observed, provide a basis for inferring population differences. In other words, the experimenter in our class-size example wants to know whether small classes will improve algebra learning for students other than those tested in his experiment. He would like to find out whether a similar effect is likely to occur in any school, with any teacher, and so on. We see, then, that the investigator wants to decide whether the two population means (e.g., of all students who might be enrolled in small classes and all who might be enrolled in large ones) differ and that he wants to make the decision on the basis of the available sample information.

Following the notation introduced earlier, let us call the imaginary population mean of final examination scores for those enrolled in small classes $\mu_S$ and the imaginary mean of those in large classes $\mu_L$. Then, in our example, we would be trying to decide between the following statements:

$$\mu_S \neq \mu_L \quad \text{and} \quad \mu_S = \mu_L$$

The researcher's task is to come to some conclusion about these statements on the basis of sample evidence from each population. In our algebra experiment, the following sample means were obtained:

$$\bar{X}_S = 80.00 \quad \text{and} \quad \bar{X}_L = 73.87$$

The difference between these sample means is certainly not 0, as might be expected if their respective population means were equal. Rather, the difference seems to be substantial (6.13 points to be exact). Can we conclude that the means of the small- and large-class populations are not equal? Does the obtained result speak for itself?

## STATES OF THE WORLD AND DECISIONS

As you can probably reason yourself by now, the obtained result does not speak for itself. We know that because of *sampling error* 80.00 is probably not the exact mean of the entire small-class population; similarly, 73.87 is probably not the exact mean of the entire large-class population. The true mean for either population may be larger or smaller than its sample mean. Therefore, it is certainly possible that $\mu_S$ does equal $\mu_L$; yet, we have obtained a difference between the two samples that is as large as 6.13 points. According to the same reasoning, $\mu_S$ may *not* equal $\mu_L$; yet, we could have obtained $\bar{X}_S = \bar{X}_L$.

In summary, we have two possible states of the world ($\mu_S = \mu_L$ is a *true* statement, or $\mu_S = \mu_L$ is a *false* statement), and the researcher's problem is to decide which one is correct.

### Null Hypothesis Testing

The decision must be one based on a logical inference, and here a special concept, *null hypothesis testing*, comes into play. The procedure and the reasoning that underlies it are as follows: First, samples of data are taken, for example, the means of two groups that have received different treatments. The null hypothesis would be that the population means do not differ, which is equivalent to saying $\mu_S - \mu_L = 0$. Then, by analyzing sample data, the experimenter determines whether the sample means differ sufficiently to reject the null hypothesis. If the null hypothesis can be rejected, then the experimenter has evidence supporting his actual hypothesis. But if the sample means do not differ sufficiently, the experimenter cannot reject the null hypothesis; more specifically, no firm decision is reached. The experimenter does not *accept* the null hypothesis; he only *fails to reject* it. We shall have more to say about this last point later.

The question of how much the sample means must differ in order for the null hypothesis to be rejected leads us to the concept of *statistical significance*. As we noted earlier, we are dealing with samples but wish to draw inferences about populations. We want to be confident, then, that observed sample differences, if obtained, actually reflect differences in the population. The critical use of statistical tests therefore becomes one of determining the likelihood that a given sample difference will occur by chance if the null hypothesis is correct.

One way to determine whether a difference between sample means occurred merely by chance would be to repeat the experiment many times; if the same difference kept showing up, we would soon be confident in rejecting the null hypothesis. Statistical tests are used because we want to do the experiment once and avoid the enormous cost of repeating it many times. The test will tell us how likely a particular mean difference would be to occur by chance; those unlikely to occur by chance are termed *significant* differences and form the basis for scientific conclusions.

But now another question arises: How unlikely must a particular difference be to be called significant? In other words, what margin of error should the experimenter be willing to tolerate? In practice, social scientists usually set this value at .05. This level of significance is commonly called the *.05 level*; it is written $p < .05$ and read "probability less than 5 percent." Sometimes, a social scientist wishes to be even more cautious about rejecting the null hypothesis and therefore chooses the .01 level of significance ($p < .01$). The likelihood or probability of *incorrectly* rejecting the null hypothesis is thus (1) a decision that must be made by the experimenter and (2) set at a quite low level of risk, usually .05 or .01. However, even with a demanding level of significance, the experimenter still runs the risk of erroneously rejecting the null hypothesis when it is, in fact, true in the population. Such errors are called *Type I* errors.

*Why the null hypothesis cannot be accepted.* Suppose a child loses the key to his bicycle lock and looks for it on the school playground. After fifteen minutes or so of looking, he returns home and announces that the key has been lost irretrievably. "Did you look in the playground?" his mother asks. "Yes," he replies, "and it isn't there." "I think it probably *is* on the playground, young man", he is admonished. "You just didn't spend enough time looking for it."

Scientists are reluctant to accept the null hypothesis because of concerns similar to those held by the mother in our example. If the boy had *found* the key on the schoolyard, we could certainly know that was where it was, but a search that fails to turn up something may simply have been too short or too weak to do so. Or the boy may simply not have looked in the right place. Similarly, an experiment that does not turn up a significant difference between two treatments may fail to do so because there were too few subjects or because the treatments were poorly carried out, did not last long enough, or any of a host of other reasons. Recognizing that the treatments could still potentially work in a better experiment,* the researcher prudently refuses to accept the null hypothesis.

*Failing to reject the null hypothesis.* Although investigators usually do not accept the null hypothesis, they do *fail to reject it*; and in certain true states of the world, this decision will be wrong. Failing to reject the null hypothesis when it is in fact false is called a *Type II* error.

Table 5-2 summarizes the distinction between Type I and Type II errors, using our class-size example. Note that two hypotheses are considered: $H_E$,

TABLE 5-2 States of the world and possible decisions in research design

| | True States of the World | |
|---|---|---|
| Decision | $H_0$ *true* | $H_0$ *false* |
| Reject $H_0$ | Type I error | Correct decision |
| Do not reject $H_0$ | Correct decision | Type II error |

$H_0$: Class size does not affect algebra scores
$H_E$: Class size does affect algebra scores

the experimental hypothesis, which states that class size does affect algebra scores and its obverse, $H_0$, the null hypothesis, which states that class size does not affect learning. The question is whether $H_E$ or $H_0$ is true in the population, and on the basis of a single experiment, the researcher must decide to reject or fail to reject $H_0$. Rejecting $H_0$ is logically equivalent to accepting $H_E$. (Note, then, that experimental hypotheses predicting effects *can* be accepted; only null hypotheses present the logical problem of acceptance versus nonrejection.)

*The idea of a better experiment is developed more fully in Chapter 6.

Thus, according to Table 5-2, there are two circumstances in which no error is made. First, $H_0$ may be correctly rejected; that is, if class size does actually influence learning in the population, we can correctly reject the null hypothesis, $H_0$, that class size does not influence learning. Second, the investigator may correctly fail to reject $H_0$ when it is actually true in the population, that is, in a world where class size does not influence learning.

There are also two circumstances in which an error is made. The upper-left-hand quadrant shows the error of rejecting the null hypothesis incorrectly; such Type I errors are of particular concern to most scientists because they may lead to putting "false facts" into the scientific literature. The probability of making a Type I error is directly controlled by the experimenter when a significance level is selected. When the .05 level of significance is used, the probability of a Type I error is 5 in 100; when the .01 level is used, the probability of a Type I error is 1 in 100. Traditionally, the probability of making a Type I error in any given experiment is designated by the lowercase Greek letter alpha ($\alpha$). Therefore, the significance level used in a particular study and its alpha level are always the same, and the terms are in fact synonymous.

The error of failing to reject $H_0$ (the null hypothesis) when it is in fact false in the population (the Type II error) is shown in the lower-right-hand quadrant of Table 5-2. The probability of a Type II error is designated by the lowercase Greek letter beta ($\beta$). $\beta$ is not set by the experimenter directly, but steps can be taken to reduce $\beta$ and thus increase an experiment's sensitivity.

In the preceding sections, we presented the basic strategy for setting up a decision rule to test a null hypothesis about two population means when only sample data are available. Of course, in order to perform the actual statistical analysis, you need to know the likelihood of various differences between sample means if the null hypothesis is true. One way to do this would be to set up a situation in which the means of two populations are known to be equal, repeatedly draw samples of size $n$ from each of them, compute the mean difference ($\bar{X}_1 - \bar{X}_2$) in each case, and plot the sampling distribution of differences that you obtained. With this sampling distribution in hand you would be able to determine the likelihood of obtaining a mean difference of any given size between your samples, by noting the area under the curve occupied by differences as large or larger than the one you obtained. Such a strategy closely parallels the technique described in Chapter 3 for determining the likelihood that a single sample mean has been drawn from a given population. However, as with our earlier example, you would not want to proceed this way in practice because of the enormous time and expense involved. It would clearly be preferable to match your situation to a known family of curves rather than having to generate a new one for each study you did. Fortunately, this is quite easy.

## THE *t*-TEST

When both populations of scores are normally distributed and their variances are equal, it can be shown mathematically that the sampling distribution of differences between two means is distributed as $t$, with $N_1 + N_2 - 2$ *df*. ($N_1$ is the number of observations in Group 1 and $N_2$ is the number of observations in Group 2.) Fortunately, even when the populations are not normal and their variances are not equal, the sampling distribution of the difference between two means tends to approximate the $t$ distribution closely (with $df = N_1 + N_2 - 2$) as long as the two samples are not markedly different in size.

In order to use the $t$ distributions, you need to express your obtained outcome as a standard score. That is, you need to determine how much, in standard-deviation units, the obtained difference between the two sample means ($\bar{X}_1 - \bar{X}_2$) deviates from the expected difference between the two population means ($\mu_1 - \mu_2$). The standard deviation of a sampling distribution of the differences between means is often called the *standard error of the difference between means* and is denoted $s_{\bar{x}_1 - \bar{x}_2}$. It follows that the formula for the standard score for an obtained difference is

$$\frac{(\bar{X}_1 - \bar{X}_2) - (\mu_1 - \mu_2)}{s_{\bar{x}_1 - \bar{x}_2}}$$

This standard score is called $t$ when the theoretical model you are using is a $t$ distribution and $z$ when the model you are using is a normal distribution. For the null hypothesis $\mu_1 = \mu_2$, the mean of the sampling distribution of differences would be 0. You now have all the information necessary for the numerator of the formula. The last piece of information you need is how to estimate the denominator, the standard deviation of the sampling distribution of differences, from the information you would ordinarily have available.

### Standard Deviation of the Difference Between Means

The standard deviation of the sampling distribution of the difference between two means can be estimated from the data from two samples according to the following formula:

$$s_{\bar{x}_1 - \bar{x}_2} = \sqrt{s^2_{\bar{x}_1} + s^2_{\bar{x}_2}}$$

This is equivalent to

$$s_{\bar{x}_1 - \bar{x}_2} = \sqrt{\frac{s_1^2}{N_1} + \frac{s_2^2}{N_2}}$$

As you can see, the standard deviation of a distribution of sample differences is very similar to the standard deviation of a distribution of sample means: It depends on the amount of variability in the original populations and on the size of the samples. The standard error of the differences between means increases with greater variability in the populations and decreases with increasing sample size. Thus, given $\mu_1 = \mu_2$, as you increase sample size, the likelihood of obtaining a large difference between $\bar{X}_1$ and $\bar{X}_2$ by chance becomes increasingly remote.

The formula should also indicate to you that the standard deviation of a distribution of the difference between two means will be larger than that for the sampling distribution of either of the means involved in the subtraction. This is illustrated in Figure 5-1.

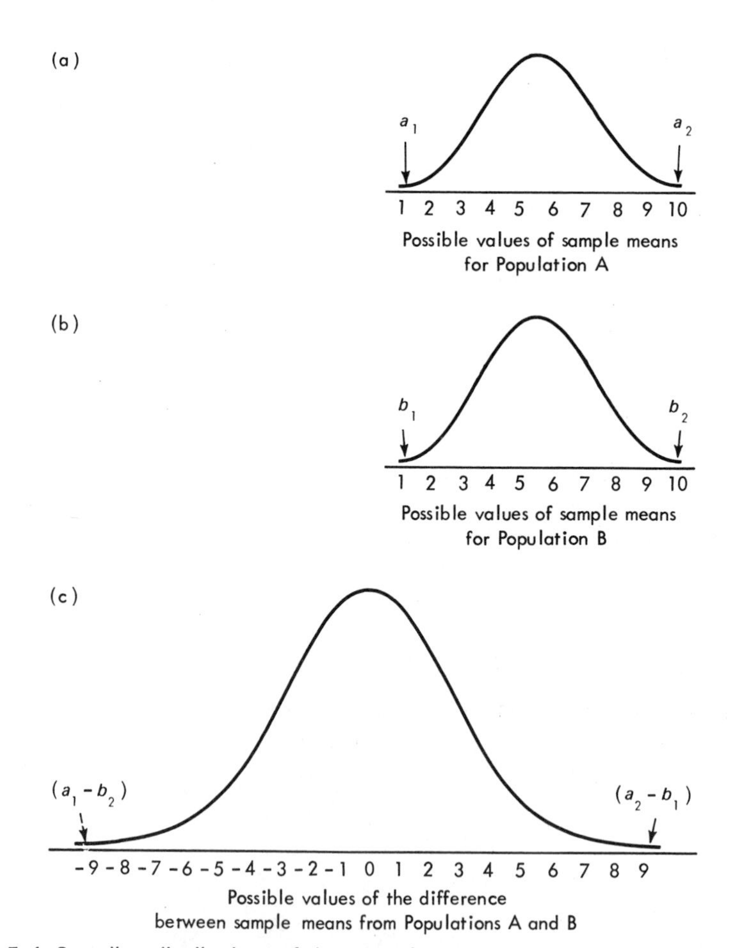


FIG. 5–1 Sampling distributions of the mean for two populations, A and B, and the sampling distribution of the difference between sample means. Notice that the range of possible differences between means is greater than the range of possible values of the mean from either population.

Suppose you had a population of scores that could range from 1 to 10. Theoretically, values for sample means from this distribution could also range from 1 to 10. Assume that you had two such populations and that you wanted to estimate the range of possible values for the difference between sample means from the two populations. The most extreme positive difference would be $a_2 - b_1 = 10 - 1 = +9$. The most extreme negative difference would be $a_1 - b_2$, or $1 - 10 = -9$. Thus, the range of possible values of the sampling distribution of differences between sample means from these two populations is 18 [$9 - (-9)$]; whereas the range of possible values for either mean alone is only 9.

*Computing the standard deviation of the sampling distribution of the difference between two means.* The following algebraic manipulations will ultimately simplify the computational procedure for obtaining $s_{\bar{x}_1 - \bar{x}_2}$:

$$s_{\bar{x}_1 - \bar{x}_2} = \sqrt{\frac{s_1^2}{N_1} + \frac{s_2^2}{N_2}}$$

$$= \sqrt{\frac{N_2 s_1^2 + N_1 s_2^2}{N_1 N_2}}$$

and if $s_1^2 = s_2^2$, then

$$= \sqrt{s^2\left(\frac{N_2 + N_1}{N_1 N_2}\right)}$$

Assuming that $\sigma_1^2 = \sigma_2^2$, $s_1^2$ and $s_2^2$ are two independent estimates of the same value. There is no particular reason to prefer one estimate to the other because each may by chance differ from the population variance. The usual practice is to *pool* the information from these two estimates on the reasonable assumption that an estimate of the population variance based on two samples is better than an estimate based on one sample.

A formula for pooling the information from two independent samples in order to obtain a pooled estimate of the population variance ($s_p^2$) is

$$s_p^2 = \frac{SS_1 + SS_2}{N_1 + N_2 - 2}$$

As you can see, the sums of squares (the basic index of variability) for each sample are added together, and this total is divided by

$$[(N_1 - 1) + (N_2 - 1)] = N_1 + N_2 - 2.$$

When $N_1 = N_2$, the $s_p^2$ calculated according to this formula is simply the average of the sample variances. When $N_1 \neq N_2$, this formula gives somewhat more weight to the estimate of variability obtained from the larger sample.

In summary, the following formulas will allow you to estimate the standard deviation of a sampling distribution of differences between means (such as that shown in Figure 5-1):

$$s_{\bar{x}_1-\bar{x}_2} = \sqrt{s_p^2\left(\frac{N_1 + N_2}{N_1 N_2}\right)}$$

where

$$s_p^2 = \frac{SS_1 + SS_2}{N_1 + N_2 - 2}$$

### Comparing Two Populations: Example 1

Imagine that you were interested in the relationship between anxiety and performance among medical school students and that you decided to compare the anxiety level of two populations: those first-year students who have failed one or more courses during the fall semester and those students who have not failed any courses their first semester. You might randomly select 12 students from each population and ask them to fill out a questionnaire designed to assess anxiety level. Possible data from this hypothetical study are given in Table 5-3a. Each student's score represents the number of high-anxiety items the student said "yes" to on the questionnaire (e.g., "I often feel that I have too many responsibilities pressing in on me").

As you can easily see from Table 5-3a, the mean anxiety score of the Failed group was higher than that of the Passed group. The difference between the two means was $\bar{X}_F - \bar{X}_P = 6.00 - 3.50 = 2.50$. In light of this evidence, should you reject the null hypothesis that $\mu_F - \mu_P = 0$?

We know that if the real difference between the population means was 0, we could still obtain a difference as large as 2.5 simply because of random fluctuations in sample means. Is our obtained difference unlikely enough to allow us to conclude that the null hypothesis is probably false? The problem is diagrammed in Figure 5-2.
As indicated, the sampling distribution of $\bar{X}_F - \bar{X}_P$ has a mean of 0 and a standard deviation of

$$(s_{\bar{x}_F-\bar{x}_P}) = \sqrt{s_p^2\left(\frac{N_F + N_P}{N_F N_P}\right)}$$

where

$$s_p^2 = \frac{SS_F + SS_P}{N_F + N_P - 2}$$

Further, $\bar{X}_F - \bar{X}_P$ is distributed as $t$ with 22 $df$

$$N_F + N_P - 2 = 12 + 12 - 2 = 22.$$

Suppose we decided to set our probability of a Type I error (our alpha level) at the conventional .05. That is, we want the probability of falsely rejecting the hypothesis $\mu_F - \mu_P = 0$ to be no more than 5 in 100. Using the table in Appendix D, we know that for $t$ with 22 *df*, a standard score either less than $-2.074$ or greater than $+2.074$ would occur no more than 5 times in 100. Now, our problem is simply to determine whether or not the standard score corresponding to our obtained difference falls into the rejection region.

Table 5-3b demonstrates a way to set up the data in order to simplify the computation of $s_{\bar{x}_F - \bar{x}_P}$. First, you need a pooled estimate of the variance in the two samples:

$$s_p^2 = \frac{SS_F + SS_P}{N_F + N_P - 2}$$

$$SS_F = \sum X^2 - \frac{(\sum X)^2}{N} \qquad SS_P = \sum X^2 - \frac{(\sum X)^2}{N}$$

$$= 486 - \frac{72^2}{12} \qquad = 194 - \frac{42^2}{12}$$

$$= 486 - \frac{5184}{12} \qquad = 194 - \frac{1764}{12}$$

$$= 486 - 432 \qquad = 194 - 147$$

$$= 54 \qquad = 47$$

$$s_p^2 = \frac{54 + 47}{12 + 12 - 2}$$

$$= \frac{101}{22}$$

$$= 4.59$$

Next, use your pooled estimate of the variance to estimate the standard deviation of the distribution of differences.

$$s_{\bar{x}_F - \bar{x}_P} = \sqrt{s_p^2\left(\frac{N_F + N_P}{N_F N_P}\right)}$$

$$= \sqrt{4.59\left(\frac{12 + 12}{(12)(12)}\right)}$$

$$= \sqrt{4.59\left(\frac{24}{144}\right)}$$

$$= \sqrt{4.59\left(\frac{1}{6}\right)}$$

$$= \sqrt{0.76}$$

$$= 0.87$$

Now, express your obtained difference as a standard score.

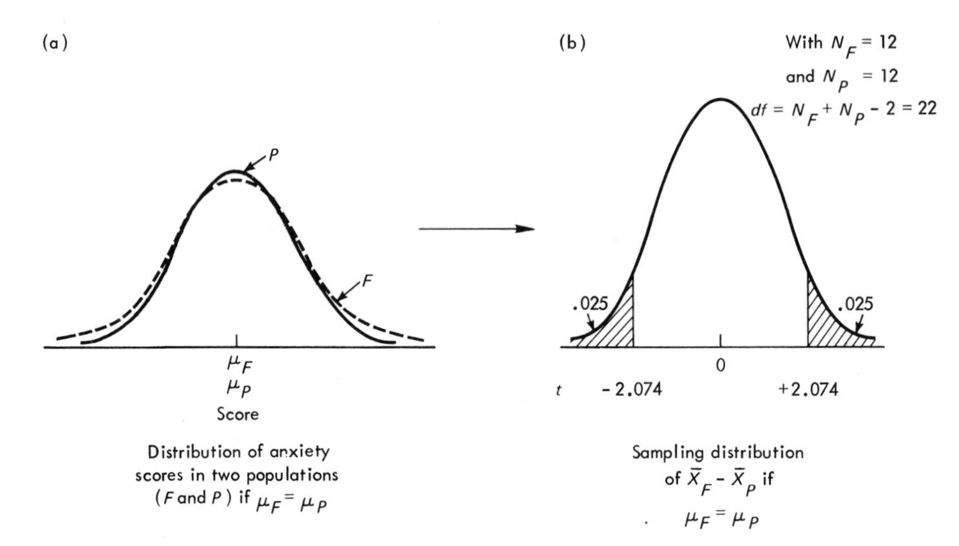


FIG. 5–2 Diagram illustrating the logic of testing the null hypothesis that $\mu_F = \mu_P$. If the null hypothesis were true and the two population variances were equal, the distributions of scores in the two populations would be identical, as shown in (a). The distribution of possible differences between sample means that could be drawn from these two populations (when $N_F = 12$ and $N_P = 12$) is shown in (b). With 22 *df* values of $\bar{X}_F - \bar{X}_P$ that yield *t* values below −2.074 or above +2.074 each should occur approximately 2.5 percent of the time (i.e., .025), and thus the overall likelihood of obtaining a *t* value this extreme by chance is less than 5 percent ($p < .05$). In such instances, the null hypothesis is traditionally rejected.

$$\text{Standard score} = \frac{\text{obtained score} - \text{mean of the distribution}}{\text{standard deviation of the distribution}}$$

$$t = \frac{(\bar{X}_F - \bar{X}_P) - (\mu_F - \mu_P)}{\sqrt{s_p^2[(N_F + N_P)/N_F N_P]}}$$

$$= \frac{2.5 - 0}{0.87}$$

$$= +2.87$$

Our obtained *t* value of +2.87 exceeds the critical value of +2.074; our sample results fall in the upper .025 of the sampling distribution of $\bar{X}_F - \bar{X}_P$, assuming that $\mu_F = \mu_P$. Thus, our sample evidence is sufficiently unlikely for us to reject the hypothesis that $\mu_F = \mu_P$.

We may infer that if we had been able to test all members of both populations, we would have found that the population of medical students who failed one or more first-semester courses would have had higher anxiety scores than the population of those who passed all their first-semester courses.

TABLE 5-3 (a) Hypothetical anxiety scores for two groups of medical school students. (b) repeats these scores and shows the preliminary calculations needed for computing *t*.

(a)

| | *F* *Students who* failed *1 or more courses* | *P* *Students who* passed *all courses* |
|---|---|---|
| | 4 | 1 |
| | 7 | 4 |
| | 1 | 2 |
| | 5 | 4 |
| | 9 | 6 |
| | 7 | 3 |
| | 4 | 5 |
| | 6 | 7 |
| | 8 | 3 |
| | 8 | 0 |
| | 7 | 5 |
| | 6 | 2 |
| $\Sigma$ | 72 | 42 |
| $\bar{X}$ | 6.00 | 3.50 |

(b)

| | *F* | | *P* | |
|---|---|---|---|---|
| | $X$ | $X^2$ | $X$ | $X^2$ |
| | 4 | 16 | 1 | 1 |
| | 7 | 49 | 4 | 16 |
| | 1 | 1 | 2 | 4 |
| | 5 | 25 | 4 | 16 |
| | 9 | 81 | 6 | 36 |
| | 7 | 49 | 3 | 9 |
| | 4 | 16 | 5 | 25 |
| | 6 | 36 | 7 | 49 |
| | 8 | 64 | 3 | 9 |
| | 8 | 64 | 0 | 0 |
| | 7 | 49 | 5 | 25 |
| | 6 | 36 | 2 | 4 |
| $\Sigma$ | 72 | 486 | 42 | 194 |
| $\bar{X}$ | 6.00 | | 3.50 | |

### Comparing Two Populations: Example 2

Imagine you are studying factors that influence the judgments about patients' chances for improvement made by beginning graduate students in clinical psychology. Specifically, you suspect that the order in which information is received about a patient will affect a clinician's prognosis for that person. As a test, imagine you performed the following experiment: Fourteen students were randomly selected from the first-, second-, and third-year graduate classes in clinical psychology at a large university. Each student received a briefing about a patient and then saw a film of a clinician interviewing the patient. After the film, each student was asked to indicate the likelihood that the patient could be released from the state hospital within five weeks by assigning a number from 0 to 10, with 0 indicating that there was no chance at all and 10 indicating absolute certainty that the patient would be well enough to release within five weeks. All students saw the same film. The written briefing was also the same for everyone except for the last paragraph. Some students read a paragraph about the patient in which he was described as "intelligent, good sense of humor, irresponsible, defensive"; whereas others read these same traits in the reverse order, "defensive, irresponsible, good sense of humor, and intelligent." Which order a particular student saw was determined by flipping a coin, with the restriction that an equal number received the two different orders of the adjectives.

The data for this experiment are given in Table 5-4. One subject was lost in the group receiving the negative traits first because the bulb on the projector blew out, and we did not have another handy. Thus, there were 7 subjects in Group $P$ and 6 subjects in Group $N$. As you can see, on the average, the subjects in the Positive-First condition rated the patient's chances as 2.69 units better than those in the Negative-First condition did

TABLE 5-4 Hypothetical ratings by clinical psychology graduate students of patients' likelihood of recovery according to whether the patients' positive or negative traits were described first

| | *Positive First* | | *Negative First* | |
|---|---|---|---|---|
| | $X$ | $X^2$ | $X$ | $X^2$ |
| | 7 | 49 | 5 | 25 |
| | 6 | 36 | 0 | 0 |
| | 4 | 16 | 3 | 9 |
| | 5 | 25 | 3 | 9 |
| | 7 | 49 | 4 | 16 |
| | 10 | 100 | 4 | 16 |
| | 2 | 4 | — | — |
| $\Sigma$ | 41 | 279 | 19 | 75 |
| $\bar{X}$ | 5.86 | | 3.17 | |

$$\bar{X}_P - \bar{X}_N = 5.86 - 3.17 = 2.69.$$

Should we conclude that the order in which the traits were described influenced the subjects' judgments about the patient's chances for recovery?

We want to test the hypothesis that $\mu_P - \mu_N = 0$; that is, the subjects were not influenced by the ordering of the traits, and the obtained difference in prognoses was only sampling error. Assuming that $\mu_P - \mu_N = 0$, the sampling distribution of the difference between means $(\bar{X}_P - \bar{X}_N)$ should be distributed as $t$ with $df = N_P + N_N - 2 = 7 + 6 - 2 = 11$. If we decide to let $\alpha = .05$, then we know, using Appendix D, that the critical value of $t$ is 2.201. That is, we need a standard score with a value less than $-2.201$ or a value greater than $+2.201$ in order to reject the null hypothesis at the .05 level. You should be able to diagram this problem yourself. In order to express our obtained difference between means as a standard score, we need to know the standard deviation of the sampling distribution of differences. This is obtained as follows:

$$SS_P = 279 - \frac{41^2}{7} \qquad SS_N = 75 - \frac{19^2}{6}$$

$$= 279 - \frac{1{,}681}{7} \qquad = 75 - \frac{361}{6}$$

$$= 279 - 240.14 \qquad = 75 - 60.17$$

$$= 38.86 \qquad = 14.83$$

$$s_p^2 = \frac{SS_P + SS_N}{N_P + N_N - 2}$$

$$= \frac{38.86 + 14.83}{7 + 6 - 2}$$

$$= \frac{53.69}{11}$$

$$= 4.88$$

then

$$s_{\bar{X}_P - \bar{X}_N} = \sqrt{s_p^2\left(\frac{N_P + N_N}{N_P N_N}\right)}$$

$$= \sqrt{4.88\left(\frac{7+6}{7 \cdot 6}\right)}$$

$$= \sqrt{4.88\left(\frac{13}{42}\right)}$$

$$= \sqrt{\frac{63.44}{42}}$$

$$= \sqrt{1.51}$$

$$= 1.23$$

Now, convert our obtained difference into a standard $t$ score.

$$
\begin{aligned}
t &= \frac{(\bar{X}_P - \bar{X}_N) - (\mu_P - \mu_N)}{\sqrt{s_p^2[(N_P + N_N)/N_P N_N]}} \\
&= \frac{2.69 - 0}{1.23} \\
&= 2.19
\end{aligned}
$$

Because our obtained $t$ of 2.19 does not fall within our region of rejection (greater than $+2.201$ or less than $-2.201$), we may not reject the null hypothesis. Our evidence is not sufficiently strong to support the conclusion that the order in which traits were described influenced the subjects' judgments.

In addition to providing practice conducting $t$-tests of hypotheses of the form $\mu_1 - \mu_2 = 0$, these last two examples also illustrate another important point first made in Chapter 1. They are examples of the differences between *experiments*, designed to yield causal statements and *nonexperimental studies*, designed to yield statements about possible differences among populations or relationships among measures. In Example 2, an experiment, the independent variable (manipulated by the experimenter) was the order of information and the dependent variable was the graduate students' judgments. The question was whether the order in which people receive information causes changes in their judgments.

In Example 1, a nonexperimental study, subjects belonged to one of two classifications ("failed one or more courses" or "passed all courses") before the investigation began. Performance in medical school is not a true independent variable because the investigator did not decide who would pass and who would fail—that is, this variable was not manipulated. Classifications or characteristics which we study but which we do not manipulate are often called *subject variables* because they come with the subjects. The measure of interest in Example 1 was each person's anxiety score, and for it we might use the more general term *response measure* rather than dependent variable. The study was designed to tell us whether anxiety level is different for people falling into these two classifications, but a statistically significant difference between them would not allow us to conclude that performance in school *causes* changes in anxiety. These two populations of people could differ in many other ways besides whether or not they failed a course, and any of these other differences may be the "real" reason anxiety scores for the two groups differed. Also, this design does not allow us to rule out the possibility that students began the term with different degrees of anxiety, which produced differences in school performance rather than vice versa. The conditions under which causal inferences can and cannot be drawn will be discussed in more detail in Chapters 6 and 11.

## Practice Problems

### *A. You Are Getting Very Sleepy*

An experimenter was interested in whether the level of illumination in a room influences the time it takes to hypnotize people. There were two treatment conditions: bright light and dim light. The subjects were randomly selected from a population of adult volunteers who had all been hypnotized at least once before. Each subject was randomly assigned to one of the conditions. The number of minutes it took to hypnotize each subject is given below:

| Bright Light | Dim Light |
|---|---|
| 1 | 3 |
| 1 | 1 |
| 5 | 2 |
| 5 | 2 |

1. What would be the Type I and Type II errors in this experiment?
2. Analyze these data.

### *B. Where the 'Gators Roam*

Six different points from which a river in Florida could be observed were randomly selected: three in the upper-river region and three in the lower-river region. The number of alligators observed during a four-hour period was recorded.

| Upper River | Lower River |
|---|---|
| 10 | 5 |
| 5 | 1 |
| 7 | 3 |

Analyze these data.

### *C. At How Many Dollars a Day?*

At a recent medical convention, ten doctors who had each used one of two alternative surgical methods reported the length of time their patients were kept in the hospital.

Reported number of days in hospital

| Old Surgical Method | New Surgical Method |
|---|---|
| 2 | 4 |
| 1 | 8 |
| 3 | 1 |
| 2 | 1 |
| 7 | 6 |

1. Determine whether patients treated by the two methods differ in length of hospitalization.
2. Discuss the conclusions that can and cannot be drawn from this result.

## *D. Jekyll and Hyde*

A teacher was interested in whether the attitude of the person administering a test would influence children's performance on conceptual problems. To one group of children (Group P), he was pleasant and encouraging during the test; to another group of children (Group C), he was cold and uninterested. The teacher assigned children to groups by taking every other name in a class roll book and placing those children in Group P; the remaining children were then assigned to Group C. The number of problems correctly solved during the test period was recorded for each child. Unfortunately, three of the children were absent on the day the experiment was conducted.

| Group P | Group C |
|---|---|
| 5 | 5 |
| 5 | 0 |
| 7 | 7 |
| 10 | 5 |
| 10 | — |
| — | — |

Analyze these data.

## *E. The Numbers Game*

Analyze the data from Table 5-1. Did class size have an effect on performance on the algebra test? The following values will simplify your calculations:

$$\sum X_S = 2{,}400 \qquad \sum X_L = 2{,}216$$

$$\sum X_S^2 = 198{,}596 \qquad \sum X_L^2 = 173{,}046$$

# Designing Powerful Experiments

# 6

AS DISCUSSED IN CHAPTER 5, to obtain acceptable support for an experimental hypothesis, social scientists must in fact produce sufficient evidence to reject the null hypothesis. This aim is most likely to be achieved when $\beta$, the probability of a Type II error, is minimized and power is therefore maintained at a sufficiently high level. In this chapter, we will consider the ingredients that make for a powerful experiment and also mention several pitfalls that render experiments either weak or confounded. And although our statistical discussion will revolve around $t$, the general principles that follow apply to much more complex statistical designs and procedures as well.

## FACTORS AFFECTING THE MAGNITUDE OF $t$

A *powerful* experiment or study is one that maximizes the chances you will detect differences between populations by rejecting the null hypothesis. As you now know, statistically this will require large values of $t$. Therefore, it will be helpful to examine the formula for a standard $t$ score in more detail in order to understand what a sensitive experiment is.

$$t = \frac{(\bar{X}_1 - \bar{X}_2) - (\mu_1 - \mu_2)}{\sqrt{s_p^2[(N_1 + N_2)/N_1N_2]}}$$

There are three critical components of this formula: (1) the difference between the two sample means, (2) the pooled estimate of the variability in the populations, and (3) the sample size. Increases in the value of $t$ will be obtained by increasing the difference between the two means ($\bar{X}_1 - \bar{X}_2$), decreasing the variability among scores ($s_p^2$), and/or increasing the sample sizes ($N_1$ and $N_2$). In any particular investigation, we would not necessarily have control over all these factors; but for the purposes of illustration, we will take a case in which we can influence each of them.

Suppose a social psychologist wanted to test the statement that the better people know each other, the more effectively they will work as a group in solving problems. Assume that, unknown to the psychologist, this statement is in fact true. Imagine an experiment in which groups of three people were given 30 minutes to solve problems. Each group's score was the number of problems solved in 30 minutes. (Notice that here each group was a subject in the sense that each group yielded one observation or score.) The people participating in the experiment were students from three high schools in a large metropolitan area; one student from each school was randomly assigned to each group so that none of the students within a particular group knew each other. Half of the groups, selected randomly, were given 5 minutes to get acquainted before the problem-solving session; the students in the other half of the groups spent the first 5 minutes chatting individually with paid research assistants and received only a brief introduction to the other members of the group prior to the problem-solving session. The data from this experiment are shown in Figure 6-1a. On the average, the students with the 5-minute Get-Acquainted period solved one more problem than the students with No-Get-Acquainted period did. Verify that the $t$ score corresponding to $\bar{X}_{GA} - \bar{X}_N = 6 - 5 = 1$ is 1.96.

The null hypothesis is $\mu_{GA} - \mu_N = 0$. With 18 $df$, the value of $t$ must be less than $-2.101$ or greater than $+2.101$ in order to reject the null hypothesis at the .05 level. Thus, these data do not permit us to reject the hypothesis that $\mu_{GA} = \mu_N$. Our evidence is not sufficient to conclude that allowing people to get acquainted beforehand facilitated problem solving. There are several ways we could have made this experiment more sensitive to real differences, that is, increased the probability that our experiment would yield a statistically significant set of results.

### Magnitude of the Difference Between Means

First, let us examine our manipulation of the independent variable. Suppose the mean problem-solving score of a population of groups with 5 minutes of interaction is exactly 1 unit higher than that of a population of groups with 0 minutes of interaction. Although this is a real difference, it is a very small one to try to detect. If the experimental hypothesis were true, contrasting 0 and 10 minutes of interaction would probably produce a larger difference between the sample means. And probably the difference between 0 and 30 minutes of interaction would be greater still. Perhaps our original manipulation was not strong enough to produce a sizable (easily detectable) difference. Figure 6-1b indicates an hypothetical experimental outcome in which everything is the same as outlined for Figure 6-1a except that the members of half the groups were given 30 minutes to interact individually with experimental assistants and the members of the other groups were given 30 minutes to interact among themselves. Nothing has changed from Figure 6-1a to Figure 6-1b except the difference between the means of the two conditions ($\bar{X}_{GA} - \bar{X}_N = 6$). However, the larger magnitude of the effect of the experimental manipulation substantially affects the value of $t$ (11.76 versus our former 1.96).

This same reasoning can be applied to nonmanipulated variables. Suppose you were interested in determining whether there was a difference between the white blood cell counts of younger and older adults. It might be reasonable to expect a larger difference between populations of twenty-five- and fifty-five-year-olds than between populations of twenty-five- and twenty-eight-year-olds.

In general, then, if you are simply interested in ascertaining (with a simple comparison of two samples) whether or not there is a relationship between two variables (such as amount of time to get acquainted and group problem solving or age and white blood cell count), then it is often a good strategy to try to examine samples that should be maximally different if the postulated relationship is real.

The expectation that choosing two widely divergent populations should yield a greater mean difference than less divergent populations is based on the assumption that the relationship between the variables of interest is *monotonic*; that is, as one variable increases, so does the other, through the entire range of possible values of each. In our problem-solving example, we would assume that the relationship between time available to get acquainted and number of problems solved looks something like that shown in Figure 6-2a.

However, the true relationship might be more like that shown in Figure 6-2b, which indicates that interaction time helps problem solving only up to a point, beyond which the number of problems solved starts to decrease (a *nonmonotonic* relationship). In this case, comparing samples from 0 and 30 minutes might very well lead us to believe that time to get acquainted and problem solving are not related. Thus, we should always be aware of the danger of arriving at a false conclusion because we have not imagined the

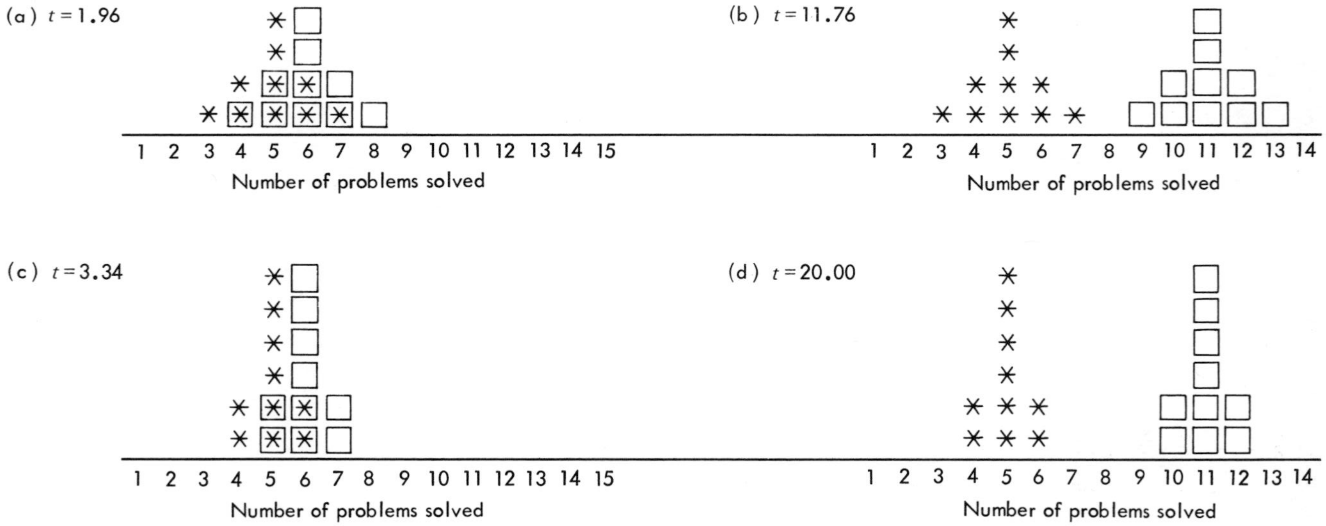

(a) $t = 1.96$

1 2 3 4 5 6 7 8 9 10 11 12 13 14 15

Number of problems solved

(b) $t = 11.76$

1 2 3 4 5 6 7 8 9 10 11 12 13 14

Number of problems solved

(c) $t = 3.34$

1 2 3 4 5 6 7 8 9 10 11 12 13 14 15

Number of problems solved

(d) $t = 20.00$

1 2 3 4 5 6 7 8 9 10 11 12 13 14

Number of problems solved

FIG. 6–1 Eight hypothetical outcomes of experiments on the effects of getting acquainted with the other members of one's group on group problem solving. In each panel, two distributions are shown: Scores designated by squares are from groups that did get acquainted before trying to solve the problems, whereas scores designated by asterisks are from groups that did not get acquainted. *t* values for the comparisons between conditions are given in each panel and illustrate that *t* values are related to the magnitude of the difference between means, the variability within each distribution, and sample size.

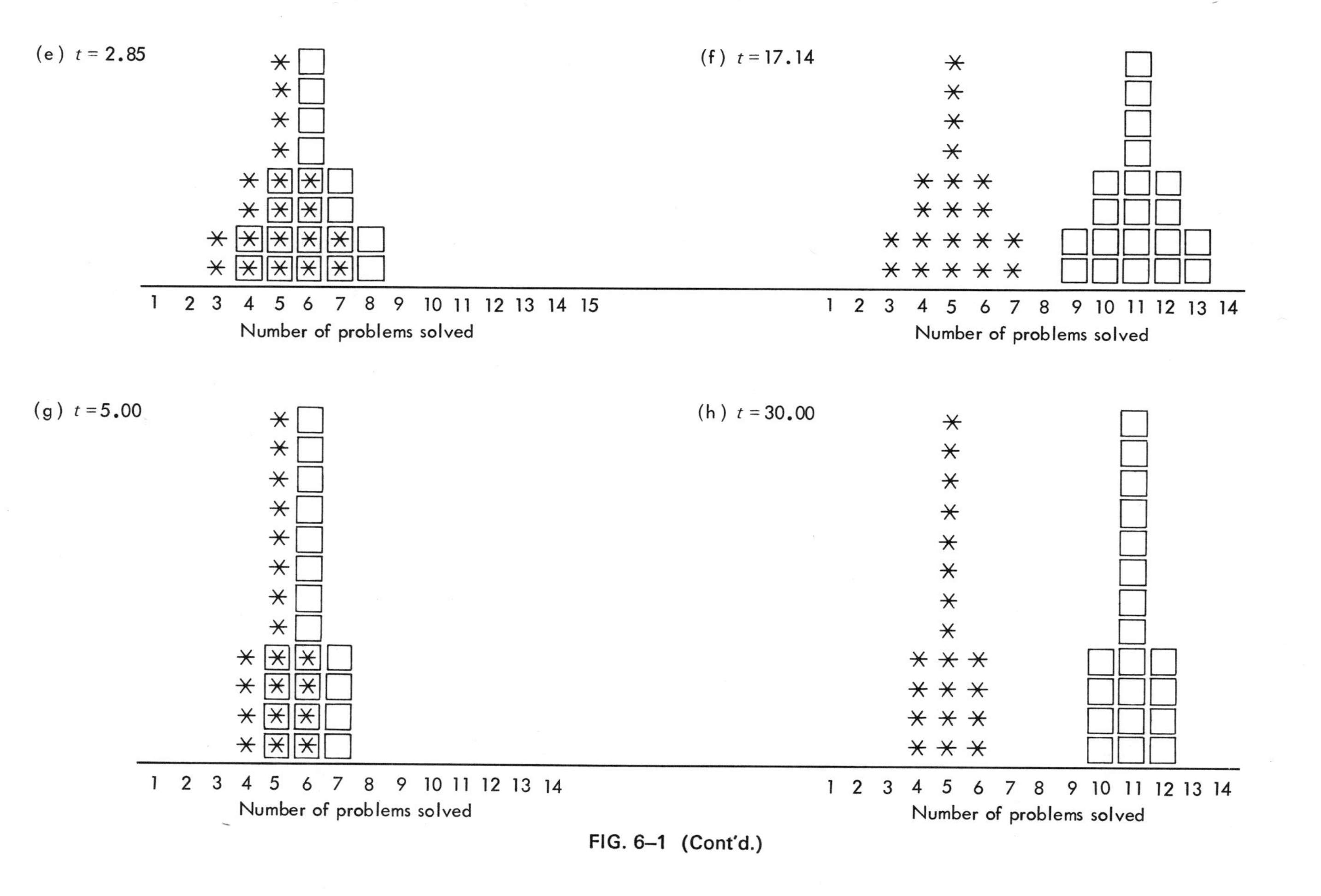
(e) $t = 2.85$
1 2 3 4 5 6 7 8 9 10 11 12 13 14 15
Number of problems solved
(f) $t = 17.14$
1 2 3 4 5 6 7 8 9 10 11 12 13 14
Number of problems solved
(g) $t = 5.00$
1 2 3 4 5 6 7 8 9 10 11 12 13 14
Number of problems solved
(h) $t = 30.00$
1 2 3 4 5 6 7 8 9 10 11 12 13 14
Number of problems solved

**FIG. 6–1 (Cont'd.)**

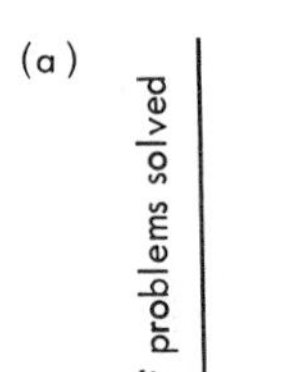
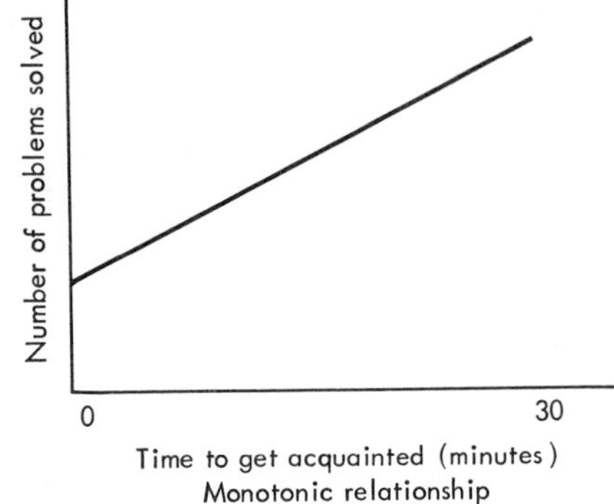


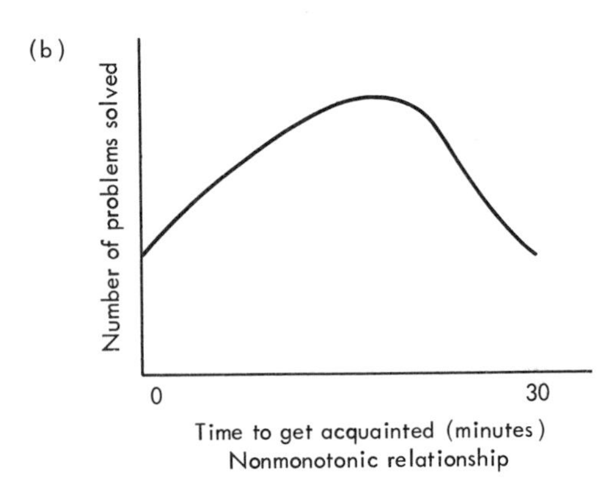


FIG. 6–2 Two hypothetical relationships between time spent getting acquainted and performance in group problem solving.

true form of a relationship—another reason never to accept the null hypothesis. But very often, selecting divergent populations is a good strategy for detecting differences.

In some cases, the strategy of trying to maximize the difference between two conditions is not possible. This is obviously true with population comparisons such as males versus females, those people who have had diphtheria versus those who have not, New Yorkers versus Californians, and so on. In addition, even in those cases in which it is possible to select more divergent populations, you might be interested in comparing less divergent ones. You might, for example, want to know specifically whether or not 5 minutes of interaction influences problem-solving ability in order to determine whether it is worth allowing this time in a management training program. In these cases, there are other things that can be done to try to maximize the chances of finding a large $t$.

### Amount of Variability in the Scores in the Population

Compare Figures 6-1a and 6-1c. The only difference between these two experimental outcomes is that there was less variability within each condition in Figure 6-1c than there was in Figure 6-1a. The practical consequence of this decrease in variability is that our computed value of $t$ increased to 3.34. Thus, in Figure 6-1c, we had a significant outcome without increasing the difference between the means of the two groups. How in fact could we decrease the variability in the scores?

To answer this question, think about the factors that produce variability among scores in the first place. Each group's problem-solving score in our example probably depends on a number of things other than simply the amount of time that the group members have had to interact. There are relatively stable characteristics of people that will influence the group's score: such things as the intelligence of the people in the group, the type of education they have had, and the amount of previous experience in problem-solving situations. There are also more transient aspects of the states of the individuals in the groups: whether or not they have had breakfast, how tired they are, whether they are worried about their schoolwork, their relationships with family or friends, their perception of the importance of the investigation, and their motivation to do well, to name a few. In addition, aspects of the environment will influence the group's score: how comfortable the room temperature is, the amount of distracting noise from outside the room, and whether or not the experimenter left out any important part of the instructions. These individual differences can contribute unwanted variability within each group. Reducing them would have made it easier to detect a significant effect in an experiment.

Some sources of within-group variability cannot be eliminated directly, of course. There is very little an investigator can do to alter the home or sex life of his subjects. But it is possible to reduce variability in a number of ways. In our example, for instance, any of the following changes might have decreased the variability in scores within the two conditions:

1. Rather than selecting randomly from tenth-, eleventh-, and twelfth-grade students, select randomly from twelfth-grade students who have successfully completed geometry. This would decrease the variability in the amount of education from one group to the next and also might reduce potential differences among groups in the problem-solving experience of the members.
2. Instead of conducting the experiment during any hour between 8:00 A.M. and 4:00 P.M., conduct the experiment only from 9:00 to 12:00 A.M. to reduce variability among individuals with respect to how tired they are. This would also eliminate sessions during hot afternoons and, consequently, decrease the variability among the scores as a result of environmental fluctuations.
3. Instead of explaining the experiment to the group members conversationally, read the instructions to them from a written script in order to minimize variations in the information each group receives.

These procedures all have the potential of decreasing the variability of the scores and thus decreasing our estimate of $s_p^2$. Because $s_p^2$ is a major contributor to our estimate of the standard deviation of the sampling distribution of differences ($s_{\bar{X}_1-\bar{X}_2}$), reducing $s_p^2$ increases values of $t$. As a general rule, then, *the less variability there is among sets of scores, the easier it is to detect small differences* in their means. In other words, the less variability among scores from a given condition, the more sensitive the experiment. However, a price must be paid. Whenever sensitivity is achieved by limiting the populations studied (e.g., only twelfth-grade students who have passed geometry rather than all tenth-, eleventh-, and twelfth-grade students), we lose something in generality inasmuch as our results permit inferences back to only those populations from which we have sampled.

### Size of the Samples

Now compare Figure 6-1a with Figure 6-1e. In the experiment pictured in Figure 6-1e, the only change is that the number of scores sampled has been increased from 10 to 20 per condition. One consequence of the larger sample is that the $t$ value of 2.85 is significant at the .05 level (with 38 *df*, the critical value of $t$ is 2.0224). In general, with all other things equal, *increasing the size of the samples increases the sensitivity of an investigation.* Thus, with a relatively large $N$, relatively smaller effects of independent variables or small differences between populations are more likely to be detected. If you expect the difference between two populations to be quite large, then fewer subjects will usually be needed. Setting $N$ at some specific value is a little like setting the resolution of a microscope or choosing the size of a fishnet; it depends on what you expect to find. As $N$ becomes larger, the estimate of the mean of each population becomes more precise, and smaller differences between populations will show up. Thus, setting $N$ is one way of determining the magnitude of the differences that are of interest to you.

An important part of research is trying to get the most information for the least cost in time and money. Thus, attention should be given to possibilities of maximizing the difference between conditions, decreasing variability (without severely limiting generality), and getting an $N$ that is appropriate for the size of differences you think are important to detect. Figure 6-1a–h summarizes the effects of these factors. Learning to make these decisions comes largely from experience in a particular field.

## "DIRECTIONAL" PREDICTIONS AND STATISTICAL POWER

Research questions involving null hypothesis testing can be either one- or two-sided, depending on whether the experimental hypothesis being tested specifies the *direction* of the expected difference. Imagine, for example, that a researcher is investigating the effects of alcohol on aggressiveness. Subjects would be assigned randomly to a Drinking (say, two ounces of alcohol) or No-Drinking condition and some measure of aggressiveness taken for each person. A $t$ test could then be performed on these sample data. And now here's the rub: The $t$ test (and the $t$ distribution), as we have described it so far, is used to determine whether the treatment had an effect *in either direction* relative to the data from control subjects. If the mean aggressiveness of the Drinking group is significantly greater than that of the No-Drinking group (falls in the right, or *upper*, .025 of the appropriate $t$ distribution) or significantly less than that of the No-Drinking group (falls in the left, or *lower*, .025 of the $t$ distribution), the null hypothesis

$$\mu_d - \mu_{nd} = 0$$

may be rejected. Thus, the total $\alpha$, or probability of a Type I error, is .05. Our experimenter specifically hypothesized, though, that alcoholic beverages would make people *more* aggressive and claims to have been penalized by a test that looked at the other alternative (that alcoholic beverages might make people *less* aggressive). If the entire $\alpha$ area (.05) had been placed in the upper tail, smaller differences between the two conditions *in the predicted direction* would be needed to reject $H_0$. Therefore, a one-tailed test, which places the tolerable probability of a Type I error on one side or "tail" of the $t$ distribution, is more powerful than a two-tailed test for detecting a difference in a predicted direction.

Generally speaking, most researchers employ statistical tests that are two-tailed, asking whether two groups differed significantly or not. But in some circumstances, it is legitimate to look for differences in one direction only. However, there is a price. *The decision to use a one-tailed test must be made a priori* (i.e., in advance of the data collection and/or based on a clear theoretical rationale), *and results in the other direction are inadmissible as evidence* (although, of course, they will influence one's future research decisions).

In sum, then, use of a one-tailed test increases the power of an investigation by an administrative decision rather than by actually changing the research, but the increased power may be enjoyed only at the price of reduced scope and generality. Decisions of this sort must be faced continually in research.

## CONDUCTING AN INVESTIGATION: ADDITIONAL CONSIDERATIONS

### Marginal Outcomes or "Borderline Significance"

Suppose that you conducted an experiment and that you obtained a $t$ value of 1.751 in a case where the critical value of $t$ at the .05 level is 2.074. According to the usual decision strategy, your result is not significant at the .05 level, and therefore you may not conclude that the experimental treatment had an effect. On the other hand, if you were willing to tolerate a Type

I error rate of 10 percent rather than 5 percent, you would be able to conclude that the experimental treatment had an effect. There is, indeed, something arbitrary about conventional alpha levels. They are useful rules of thumb, but that is all. Ordinarily, and quite reasonably, an investigator is influenced by the magnitude of the obtained value of $t$, not just simply by whether or not it exceeds a critical value. Investigators often talk with more confidence about results that fall in the extreme .01 of a sampling distribution than they do about results that fall in the extreme .05 of the distribution. Similarly, when a result is "almost" significant, an investigator tends to believe that a new study, perhaps with a more sensitive design, will yield a significant outcome. Sometimes, results that fall between the .10 and .05 levels of significance are described as *tendencies* in the data.*

Clearly, conventional decision procedures have been introduced to help guide thinking in situations where there is room for error and to help direct future research. Like most conventions, they help make judgments easier but do not really eliminate the need to exercise judgment. A computationally accurate statistical analysis indicating that two sets of scores are or are not from similar populations is only a small part of well-done research. The conclusions drawn from sets of numbers are only as good as the reasoning and methods that have gone into collecting them.

### Avoiding Systematic Bias

The aim of an investigation in which two populations are compared is to obtain information about possible differences among subjects with respect to some variable of interest (e.g., order of adjectives in a description, amount of problem- solving experience, or age or sex of participant). When the groups differ in some way in addition to the way the investigator intends for them to differ, a *confound* is present, and the investigation is *internally invalid.* Many research practices have been adopted in order to reduce the likelihood of confounds and secure valid data.

*Differences in the general measurement circumstances.* Subjects are generally tested in a manner designed to reduce the possibility that uncontrolled environmental factors tend to favor one group over another. For example, sometimes it is possible to run all subjects simultaneously in a single room. If so, you would want to make sure that the subjects were randomly assigned to the various chairs in the room so that, on the average, lighting, acoustic factors, and such were equally good for all treatment conditions or comparison groups.

When each subject is seen individually, how should the investigation be conducted? Suppose you ran all the subjects in one condition of your study

*They are also mistakenly called *trends*, but that term is a technical one with quite a different meaning.

and then all the subjects in the second condition. You might be less experienced and therefore make more errors in your collection of data with the first group than with the second. Or suppose some event, such as a national election, occurs that might influence your subjects' concentration. If all the subjects in one condition are run while the returns of a national election are coming in and the subjects from the other condition are not run until weeks later, you might expect the performance of the two groups to differ for reasons unrelated to the variable under investigation.

To guard against systematic biases such as these, investigations are often conducted in *randomized replications*, or *blocks*, of the various treatment conditions. If there are six conditions in the investigation, the first six subjects are assigned randomly so that there is one in each condition. The order in which they are tested is determined randomly. These six subjects constitute the first replication, or block, of the study. Then the next six subjects are assigned and tested in a similar fashion, and they constitute the second replication, or block, of the study. The process is repeated until all subjects have been run. This procedure has the advantage of allowing the influence of external and environmental factors the opportunity to fall equally on all the conditions. If you use it, then an approximately equal number of subjects in each condition will automatically be run at any given time of day; an approximately equal number of subjects in each condition will be run when you are happy and friendly and when you are irritated and abrupt; any deterioration of the equipment that might affect the accuracy of the measures will be distributed equally across conditions; and so forth.

For similar reasons, when two or more experimenters test subjects for the same investigation, each experimenter should run an equal number of subjects in each condition so that potential effects on subjects attributable to the particular experimenter are equally distributed across the conditions.

*Subject differences.* In an experiment, it must be presumed that the groups are equivalent prior to the introduction of the experimental treatments. Initial equivalence is the only circumstance that allows the causal inferences experiments are designed to yield. Subjects are therefore generally assigned randomly to treatment conditions so that, on the average, individuals in all conditions will be about equal in all abilities and/or characteristics that might contribute to their scores.

In addition, if subjects drop out of or are lost to an experiment or some other type of study, it is important that this *subject loss* is not related to the experimental treatments. Suppose, for example, you were comparing two weight-reducing programs. Subjects are randomly assigned either to Program A or to Program B. If Program A is truly more effective than Program B, Program B will include more subjects who do not lose much weight. If some of the subjects who are not losing much weight get discouraged and drop out of the experiment, more subjects may be lost from Program B than from

Program A by the end of the treatment period. If weight loss is averaged over the subjects who complete the program, the net result may be an artificially high estimate of the effectiveness of Program B and, consequently, an erroneous conclusion about the relative effectiveness of the two programs (see Table 6-1).

TABLE 6-1 Example illustrating consequence of differential subject loss on means of two groups

| | Weight-reducing Program A | Weight-reducing Program B |
|---|---|---|
| | *Mean Pounds Lost in Five Months If All People Remained in the Study* | |
| | 10 | 5 |
| | 12 | 11 |
| | 13 | 7 |
| | 8 | 10 |
| | 12 | 12 |
| $\sum X$ | 55 | 45 |
| $\bar{X}$ | 11.00 | 9.00 |
| | *Suppose Two People Who Are Not Losing Much Weight* Drop Out *of Program B* | |
| | 10 | 11 |
| | 12 | 10 |
| | 13 | 12 |
| | 8 | — |
| | 12 | — |
| $\sum X$ | 55 | 33 |
| $\bar{X}$ | 11.00 | 11.00 |

However, if subjects are lost for reasons that we can assume operate randomly across the conditions of the investigation (such as the burned-out projector bulb in Example 2, Chapter 5, page 70), then losing a few subjects should not appreciably affect the outcome of our investigation.

*"Hidden" or "unwanted" treatments.* Experimental variables often consist of a number of components, only one of which is of real interest. For example, suppose you are investigating the effect of a drug on the activity level of white mice. All mice are placed on a checkered grid, and you record the number of different squares each mouse steps on during a five-minute observation period. Prior to this, a random half of the mice have been injected with the drug and the other half have not. What factors, other than the presence

or absence of the drug, might contribute to the animals' performance in this situation? The needle prick from the injection itself might influence activity level. Or the additional time the animals were handled by the experimenter while they were injected might influence activity level. Therefore, in order to hold these other factors constant across the two conditions, you might also inject the second group of animals, but with a chemically inert solution. Thus, animals in both conditions would be subjected to the needle prick and would be handled approximately equally. Now the only difference between conditions would be the presence or absence of the drug.

*Experimenter bias.* It is sometimes possible for an investigator's expectations to influence the outcome of a study. This can happen in at least two ways: (1) The investigator can actually influence the subjects' performance. (2) The experimenter can systematically make errors in his observations. For example, the hypothetical investigator in the preceding experiment might expect greater activity from the mice receiving the drug and might be more agitated or nervous in handling the mice in the Drug condition. If he tends to treat the mice in one group differently from the mice in the other, this difference becomes an unwanted or hidden treatment. Or because a certain amount of judgment must be exercised in determining whether an animal has entered a new square or not, the experimenter might unintentionally have a less exacting criterion for animals in the Drug condition than for animals in the other condition. Both types of bias could operate despite the best efforts of the investigator to guard against them. Hence, when there is reason to expect that experimenter bias might influence the results, an investigation is often run *blind.* In the present example, the experiment could be run blind if a second experimenter coded the hypodermic needles so that the person handling the animals could not tell to which condition any particular animal belonged.

## THE PROBLEM OF GENERALITY

Someone interested in mental retardation is likely to sample subjects from a nearby hospital; someone interested in language acquisition is likely to sample children from a local school; someone interested in conformity is likely to sample students from a local college. In addition, the subjects may be run under a limited set of circumstances, for example, only by one experimenter and only between 3:00 and 5:00 in the afternoon. As we have said, technically you may generalize results only to the population from which you have sampled. Therefore, how can science proceed when the sampling is done from such relatively restricted populations of subjects, tested in such a restricted sample of situations?

The answer depends upon a willingness to assume that a limited population is characteristic or representative of a larger population. In turn, such an assumption depends upon your expectation about the degree of variability in the population to which you wish to generalize. If you assume that all people's stomachs work according to the same general principles, regardless of the particular ethnic origins and/or life experiences of the person, then you would probably be willing to generalize from the results of a study of digestive processes of a small group of fourteen-year-olds to people of all ages. However, if you have reason to believe that there is a lot of variability among members of a population with respect to some critical characteristics or experiences, then you would be more cautious in generalizing beyond the immediate data.

Often, it is quite reasonable to assume that the local population is characteristic of a much larger population with respect to a particular variable of interest. In addition, as basic findings are replicated in various laboratories around the country and the world, they begin to assume the status of generally true facts. In practice, more generality is probably obtained via replication of an outcome from successive investigations than from a single study in which the sample is randomly selected from an extremely large population.

## Practice Problems

### *A. Sigmund Who?*

Senior psychology majors who have had Introductory Psychology as freshmen or sophomores are given an alternate form of their finals. The mean score is 50 percent. The investigator asserts that because 50 percent of the material is lost in two to three years, the course has little lasting effect and should be discontinued. Do you agree with this conclusion? Why or why not?

### *B. Losing Your Better Half*

A sociologist compared measures of self-esteem of divorced and married people. There was a significantly lower mean for the divorced subjects, and the investigator concluded that failure in marriage is destructive of feelings of self-worth. Would you accept this conclusion?

### *C. Mum's the Word*

An investigator wanted to compare two truth drugs. Twenty subjects are randomly assigned to two independent groups. The subjects in Group 1 read a story, were told not to reveal any details, were given Drug A, and then were questioned by Doctor Smith. The subjects in Group 2 read the same story, were told not to reveal any details, were given Drug B, and then were questioned by Doctor Jones. The mean number of details divulged by subjects receiving Drug A was 3, and the mean divulged by subjects receiving Drug B was 10. The investigator concluded that both drugs were effective (because both groups revealed some details) and that Drug B was more effective than Drug A. What do you think?

## *D. Now You See It, Now You Don't*

A researcher is interested in what type of print allows for the easiest reading. The first 20 subjects who arrive are assigned to the Big-Print condition, and the second 20 subjects are assigned to the Small-Print condition. All subjects are run simultaneously. All subjects read the same passage and are allowed an equal amount of time. The number of words read in five minutes is recorded for each subject.

1. What is the most obvious source of potential internal invalidity in this experiment?
2. How would you redesign this experiment in order to avoid the problem you have just identified?

## *E. One for the Road*

An investigator was interested in testing the assertion that caffeine acts as a stimulant and increases alertness. The subjects were college students, and each subject was randomly assigned to one of two conditions: caffeine or no caffeine. Each subject in the No-Caffeine condition participated in the following procedure: The subject was told to watch a screen carefully; when a red light appeared, he was to press a button as fast as he could. The time to respond was recorded for each subject. Those in the Caffeine condition were run in exactly the same procedure except that before the test each subject was given a cup of coffee (the participants did not know that the coffee was related to the experiment). The experiment was conducted in randomized blocks of the experimental treatments. The data are given below in milliseconds:

| No Caffeine | Caffeine |
|---|---|
| 4 | 1 |
| 1 | 2 |
| 1 | 2 |
| 1 | 1 |
| 1 | 4 |
| 10 | 2 |

1. Analyze these data, being prepared to indicate your justification for either a one- or a two-tailed test.
2. List as many ways as you can to *improve* this experiment in order to decrease the probability of a Type II error. Be specific, and briefly indicate why each might be an improvement and yield a more sensitive experiment.
3. Suppose the experimenter had found that the Caffeine condition responded significantly faster than did the No-Caffeine condition. Give four plausible alternative explanations of such a result.

4. Do the points you raised in (3) invalidate the statistical analysis you performed in (1)?

## *F. The Numbers Game*

Look again at Practice Problem 5-D (page 74). Under what circumstances would you be particularly worried about the three children lost from the study? What assumptions would lead you *not* to worry?

# Repeated-Measures and Matched-Pairs Designs

# 7

THE STATISTICAL TECHNIQUES PRESENTED IN CHAPTER 5 allowed us to compare two sets of scores obtained from independent groups of individuals. Each score in the analysis was selected randomly and independently of every other score. Each individual participated in only one condition, and the probability of observing one person in no way influenced the probability of observing another. Sometimes, however, we want to compare two sets of scores that are *not independent* of one another. For example, we might want to compare a person's performance under one condition or treatment with his or her *own* performance under another condition. Or we might want to compare scores of pairs of subjects: husbands and wives, identical twins, littermates, or any other individuals who can be matched in some meaningful way. In such cases, we select random *pairs* of scores from a population, rather than random individual scores. And, of course, the statistical procedures we choose must take this fact into account.

## SIMPLE REPEATED-MEASURES DESIGNS

Imagine that you have been conducting a series of experiments on maze learning of rats. The animals are performing extraordinarily well, and you begin to suspect that your research assistant is somehow influencing their behavior, perhaps by inadvertently providing cues to the right responses. To test this hunch, you might randomly assign animals to each of two treatment conditions. In both conditions, each animal would receive six trials on the maze problem and would be assigned a score based on the number of trials on which it made the correct turn. In one condition, the animals would run the maze with your assistant clearly in view (the Present condition); in the other, the assistant would stand out of sight (the Absent condition). These data could then be analyzed with the independent-groups $t$ test presented in Chapter 5, testing the hypothesis $\mu_P - \mu_A = 0$. Rejection of the null hypothesis would constitute evidence that seeing your assistant during performance did indeed affect the number of correct responses made by the subjects.

There is another way you could set up an experiment to test the same hypothesis. Each rat could receive, say, twelve trials; six of them could be run with the assistant present and six with the assistant absent. Then you could compare the performance for each animal under the two conditions and get a direct look at the effects of the independent variable. Such a design is called a *repeated-measures* design because repeated measures of the same individual are taken under different conditions. (Sometimes this setup is also called a *within-subjects* or *nonindependent-groups* design.) The critical difference between this and the independent-groups design is that in the repeated-measures design each subject is compared with himself under at least two conditions and thus serves as his own control.

An hypothetical set of scores for our repeated-measures design is given in Table 7-1. Each animal is assigned a subject number, and its scores under each of the two conditions are listed in corresponding positions in the first two columns of the table.

Our reasoning in this situation is as follows: If the assistant is not affecting the animals' performance, then, on the average, the *difference* between any particular rat's scores under the two conditions should be 0. Of course, sometimes by chance (as a consequence of a momentary distraction, fluctuations in physiological state, and so on), an animal's score will be higher under one condition or the other, but overall, higher scores for one condition should be balanced by higher scores for the other. Therefore, the expected mean of the difference scores is 0 ($H_0: \mu_{P-A} = 0$). However, if performance *is* affected by the presence or absence of the assistant, then we would expect an overall difference ($H_E: \mu_{P-A} \neq 0$). The difference score ($D$) for each animal is given in the third column of Table 7-1, and the mean of the difference scores

TABLE 7-1 Mean number of correct responses by 16 rats observed as a function of whether the experimenter's assistant was present or absent in a within-subjects (repeated-measures) design. Difference scores (D) are obtained by subtracting each subject's Absent score from its Present score

| | Subject | Present | Absent | Difference $(P - A)$ |
|---|---|---|---|---|
| | 1 | 3 | 2 | 1 |
| | 2 | 4 | 3 | 1 |
| | 3 | 5 | 6 | −1 |
| | 4 | 4 | 2 | 2 |
| | 5 | 5 | 4 | 1 |
| | 6 | 5 | 3 | 2 |
| | 7 | 5 | 6 | −1 |
| | 8 | 6 | 5 | 1 |
| | 9 | 4 | 3 | 1 |
| | 10 | 2 | 3 | −1 |
| | 11 | 4 | 3 | 1 |
| | 12 | 3 | 5 | −2 |
| | 13 | 4 | 1 | 3 |
| | 14 | 5 | 4 | 1 |
| | 15 | 2 | 1 | 1 |
| | 16 | 4 | 2 | 2 |
| $\sum D$ | | | | 12.00 |
| $\bar{D}$ | | | | 0.75 |
| $\sum D^2$ | | | | 36.00 |

$(\bar{D} = 0.75)$ appears at the bottom of the column. Thus, our statistical question is whether 0.75 is sufficiently different from 0 to reject the null hypothesis.

In order to answer this question, we need to know the sampling distribution for the mean of a set of difference scores based on a sample of 16 scores. With what you already know, this is not hard to determine. Each of the difference scores can be treated like any other raw score that has been randomly and independently sampled from a population. Thus, we can think of a large hypothetical population of difference scores from which we have randomly selected 16. The sampling distribution of the mean, for samples of size $N$, was described in Chapters 3 and 4. For the present example, $\bar{D}_{P-A}$ is distributed as $t$ with $N - 1$, or 15, $df$. The mean of the sampling distribution is 0, and the standard deviation (the standard error of the mean) can be estimated from our sample data $(s_{\bar{D}} = s/\sqrt{N})$. The extent to which our obtained mean differed from what we would expect according to the null hypothesis in standard-deviation units (a $t$ score) is

$$t = \frac{0.75 - 0}{s/\sqrt{N}} \quad \text{or, in general, } t = \frac{\bar{D} - \mu_{\bar{D}}}{s_{\bar{D}}}$$

This obtained $t$ score can be compared with those listed in Appendix $D$ in order to determine whether it is sufficiently unlikely (a probability of less than .05) to lead us to reject the hypothesis that $\mu_{P-A} = 0$.

The calculations are as follows:

First

$$SS = \sum X^2 - \frac{(\sum X)^2}{N}$$

Or because in this case we have called each score $D$

$$SS = \sum D^2 - \frac{(\sum D)^2}{N}$$

$$= 1^2 + 1^2 + (-1)^2 + \cdots + 1^2 + 2^2 - \frac{(1 + 1 - 1 \cdots + 1 + 2)^2}{16}$$

$$= 36 - \frac{12^2}{16}$$

$$= 36 - 9$$

$$= 27$$

$$s^2 = \frac{SS}{N - 1}$$

$$= \frac{27}{15}$$

$$= 1.80$$

$$s = \sqrt{s^2}$$

$$= \sqrt{1.80}$$

$$= 1.34$$

$$s_{\bar{D}} = \frac{s}{\sqrt{N}}$$

$$= \frac{1.34}{\sqrt{16}}$$

$$= \frac{1.34}{4}$$

$$= 0.34$$

$$t = \frac{\bar{D}_{P-A} - \mu_{P-A}}{s_{\bar{D}}}$$

$$= \frac{0.75 - 0}{0.34}$$

$$= 2.20$$

According to Appendix *D*, we need a value larger than 2.131 to reject the null hypothesis at the .05 level; our obtained $t$ exceeds this value. We may therefore conclude that the treatment affected performance; specifically, we may conclude that rats did perform better in the presence of the assistant.

## MATCHED-PAIRS DESIGNS

Sometimes it is impractical to have each individual serve as his own control in a study. However, you can approximate this situation by finding or creating pairs of individuals who are very similar to each other. For example, suppose you were interested in comparing the effectiveness of two diet programs: the banana diet and the tomato diet. You might select participants for this experiment so that you have a set of pairs of people matched as closely as possible for age, general health, and weight prior to the introduction of the experimental programs. One member of each matched pair would be randomly assigned to Program B, and the other member would then automatically be assigned to Program T. An hypothetical set of data for this experiment is given in Table 7-2. The values in the first two columns represent the number of pounds lost by each person during the program. Scores for pair members are in corresponding positions in the table.

TABLE 7-2 Number of pounds lost under two diet programs

| Pair | Tomato Diet | Banana Diet | $D$ $(T - B)$ |
|---|---|---|---|
| 1 | 6 | 1 | 5 |
| 2 | 16 | 10 | 6 |
| 3 | 10 | 5 | 5 |
| 4 | 14 | 15 | −1 |
| 5 | 24 | 20 | 4 |
| 6 | 8 | 3 | 5 |
| $\sum D$ | | | 24.00 |
| $\bar{D}$ | | | 4.00 |
| $\sum D^2$ | | | 128.00 |

As in the repeated-measures design, a set of difference scores is obtained. The number of pounds lost by a person in Group B is subtracted from the number of pounds lost by the corresponding member of the pair assigned to Group T. These difference scores appear in the third column of Table 7-2. The computational procedures for the statistical analysis are exactly the same as those in the previous section. Note that $N =$ number of pairs.

$$SS = D^2 - \frac{(\sum D)^2}{N}$$
$$= 128 - \frac{24^2}{6}$$
$$= 128 - 96$$
$$= 32$$

$$s^2 = \frac{SS}{N-1}$$
$$= \frac{32}{5}$$
$$= 6.40$$

$$s = \sqrt{s^2}$$
$$= \sqrt{6.40}$$
$$= 2.53$$

$$s_{\bar{D}} = \frac{s}{\sqrt{N}}$$
$$= \frac{2.53}{2.45}$$
$$= 1.03$$

$$t = \frac{\bar{D} - \mu_{\bar{D}}}{s_{\bar{D}}}$$
$$= \frac{4}{1.03}$$
$$= 3.88$$

According to Appendix *D*, the critical value of $t$ with 5 *df* is 2.571. Because 3.88 is greater than 2.571, we may reject $H_0: \mu_{T-B} = 0$ and conclude that the tomato diet is superior to the banana diet.

## POWER OF THE REPEATED-MEASURES AND MATCHED-PAIRS DESIGNS

Recall that in Chapter 6 we discussed the general problem of designing sensitive studies. We said that a powerful experiment is one which maximizes the likelihood that differences between treatments will be detected. Because a null hypothesis may be rejected when the size of an obtained difference between groups is large relative to the variability among the scores, any factor that reduces variability within scores (such as increasing the homogeneity of the sample or standardizing test conditions) increases the power of an experiment. Repeated-measures and within-subjects designs generally are more sensitive or powerful than independent-groups designs because they reduce the variability of the scores with which we are working. Therefore, such designs are more likely than independent-groups designs to pick up small differences between treatments. The point can be illustrated by examining our example more closely.

Look at the scores in Table 7-2. Within the B condition, the scores range from 1 to 20; within the T condition, they range from 6 to 24. If these two sets of scores had been obtained from an independent-groups design, the resulting $t$ value would have been 0.90. With 10 *df*, our obtained value clearly would not have exceeded the critical value of 2.228. (As a review, you might verify this by working through the calculations.) However, the *difference* scores between matched pairs range only from $-1$ to 6. Why are the difference scores less variable than the scores within either of the two conditions?

A person who is very overweight may have a harder time initially losing pounds than a person who is only moderately overweight. Similarly, a person who is active and in good health may lose more easily than one who is relatively inactive or in poor health. This potential variability among the subjects will remain unaccounted for in an independent-groups design. But in a matched-pairs design, in which we would match individuals on these relevant factors in advance and take a difference score, unwanted subject variability is brought under at least partial control. Such control may greatly increase the power and sensitivity of our statistical tests.

As another example, consider the problem of evaluating two brands of automobile tires. You could simply assign a tire of each kind to motorists randomly, measure the amount of tread lost after a fixed period of driving, and evaluate the obtained difference between means with an independent-groups $t$ test. However, a moment's reflection suggests that driving habits, the make of the automobile, and road conditions will contribute a great deal to the wear on tires and that small differences in the quality of the tires themselves may go undetected given the variability attributable to these other sources. A more powerful design would be to assign one tire of each type to the rear axle of each car. This would create a set of matched pairs of scores. Each pair (consisting of one Type A tire and one Type B tire) would be subjected to the same roads, driving habits, and wheel-alignment conditions. The *difference* in tread wear for each pair should, therefore, be a relatively pure measure of difference in the quality of the tires.

Using each subject as its own control, as in the repeated-measures design, is the clearest example of the same reasoning. Differences in such factors as the learning ability, motivation, health, or distractibility of the rats in our earlier example could produce variability in scores and thereby cover up the real effects of the research assistant's presence. These effects will be held reasonably constant when each animal is measured in both conditions.

### Special Problems of Repeated-Measures Designs

In Chapter 6, we spoke of the need to avoid systematic bias in setting up and conducting a study. That same concern applies equally to the repeated-measures and matched-pairs designs. Unfortunately, the repeated-measures design also presents several additional sources of confounding that the wary investigator must avoid. One important principle is: *If each subject is to be measured in both conditions, all subjects must not be given the treatments in the same order*. The reason for this rule is that there are usually some persisting consequences, called *carry-over effects*, after any experience. Uncontrolled carry-over effects can utterly contaminate an experiment.

Suppose you were comparing people's reaction times when pressing a foot lever in response to two colored-light signals and wished to know whether responses were quicker to one color than to the other. You would very likely choose a repeated-measures design in this situation because effects produced by the color of the light would be expected to be quite small relative to individual differences in reaction time. People will vary in alertness, coordination, motivation, and a variety of other factors that could make them fast or slow. These variations, if uncontrolled, might completely mask the effects of the lights. Probably, you would also want each person to receive a number of trials with each light (say, 30) and use the average of these trials under each color condition as the person's score in that condition. Averaging over many trials in this way helps to minimize the influence of chance factors (such as temporary distraction) on the scores and therefore further reduces unwanted variability.

Suppose, now, that each person in this study first received his or her 30 trials in the Red-Light condition and then received 30 trials in the Orange-Light condition. Would this procedure be acceptable?

No. Very likely individuals would improve as they became more familiar with the procedure, more comfortable with the experimental setting, and more practiced in the task. A general improvement resulting from experience with a task or experimental procedure, usually called a *practice effect*, must not be confounded with the color conditions. Everybody might be expected to respond a bit faster in the second 30 trials because they are more practiced, but we would not want to attribute this improvement to a difference in the color of the light. The general problem is therefore to avoid confounding the order in which treatments are presented with the treatments themselves. Table 7-3 illustrates three potential consequences of failing to do so.

To simplify the discussion, we are assuming that the practice effect would be the same for everybody and would decrease reaction time approximately

TABLE 7-3 Some potential consequences of confounding the order of conditions with the conditions themselves. Data are reaction times in seconds, and we have supposed that practice decreases reaction time by 0.20 for every subject

(a)

*Suppose, in Fact,* $\mu_R = \mu_O$

| *Subject* | *Red* | *Orange (Real)* | *Real Orange plus Practice Effect* |
|---|---|---|---|
| 1 | 0.50 | 0.50 | 0.30 |
| 2 | 0.75 | 0.75 | 0.55 |
| 3 | 1.00 | 1.00 | 0.80 |
| $\bar{X}$ | 0.75 | 0.75 | 0.55 |

(b)

*Suppose, in Fact,* $\mu_R < \mu_O$

| *Subject* | *Red* | *Orange (Real)* | *Real Orange plus Practice Effect* |
|---|---|---|---|
| 1 | 0.50 | 0.70 | 0.50 |
| 2 | 0.75 | 0.95 | 0.75 |
| 3 | 1.00 | 1.20 | 1.00 |
| $\bar{X}$ | 0.75 | 0.95 | 0.75 |

(c)

*Suppose, in Fact,* $\mu_R > \mu_O$

| *Subject* | *Red* | *Orange (Real)* | *Real Orange plus Practice Effect* |
|---|---|---|---|
| 1 | 0.70 | 0.50 | 0.30 |
| 2 | 0.95 | 0.75 | 0.55 |
| 3 | 1.20 | 1.00 | 0.80 |
| $\bar{X}$ | 0.95 | 0.75 | 0.55 |

0.20 seconds. Table 7-3a is a case in which the true difference between conditions is 0; that is, the color of the light does not affect reaction time. However, the practice effect makes it appear as if reaction time is faster to the orange light than to the red one (0.55 versus 0.75 seconds). In Table 7-3b, a potential real difference, in which reaction times to the red light are faster, is actually masked by the practice effect. And finally, in Table 7-3c, a potential real difference, in which reaction times to the orange light are faster, is magnified by the practice effect.

A very similar argument can be made for contamination of the data by a *fatigue effect*. People may get tired or bored during the course of a challenging or long experimental procedure, and if one condition always follows another, this fatigue effect will contribute more to your estimate of the performance under one condition than to your estimate of the performance under the other. In the colored-light example, a fatigue effect would produce a systematic tendency to overestimate reaction time for the Orange-Light condition.

Practice and fatigue are fairly common processes, but there are numerous other carry-over effects that may also threaten particular types of studies. When comparing two drugs, for example, it would be important to be sure that the first drug administered is not still active when the second is given. In actual research situations, we can rarely specify all the possible carry-over effects, and it is virtually impossible to assess the relative contribution to our scores of carry-over effects (such as practice and fatigue) that exert their influence in opposite ways. And ordinarily, of course, we do not know the true difference between conditions.

### Counterbalancing

To guard against potential erroneous conclusions as a result of carry-over effects, a procedure called *counterbalancing* is commonly used. The general purpose of counterbalancing is very much like that of assigning subjects randomly to treatment conditions: Potential error resulting from carry-over effects is distributed equally across all conditions of the study, thereby tending to cancel out these interferences.

When subjects are run in a counterbalanced way, an equal number receive the treatments in A-B order and in B-A order. All the problems illustrated in Table 7-3 might have been avoided if half the subjects had seen the lights in the order orange-red and the other half had seen them in the order red-orange.

A complication arises when more than two measures are obtained for each subject because strict counterbalancing becomes more difficult. For example, although there are only two possible orders of two treatments (A-B and B-A), there are *six* possible orders of three treatments (A-B-C, B-C-A, C-A-B, C-B-A, B-A-C and A-C-B) and *twenty-four* possible orders of four treatments, too many to list here. One solution to this problem is to pick a few of the possible orders randomly and then, again randomly, assign subjects in equal numbers to each of the selected orders. Another alternative is to create a randomized order of treatments for each subject. In either case, by using the powerful tool of random assignment, it is possible virtually to eliminate many carry-over effects.

But a word of caution is needed. Counterbalancing the order of administration of treatments across subjects is an effective way to deal with carry-over effects only when we are reasonably certain that the carry-over effect from Treatment A to Treatment B is the same as the carry-over from Treatment B to Treatment A. For example, if one treatment is more fatiguing than the other, counterbalancing the order of presentation will not equally distribute the carry-over effects to the two conditions.

### Choosing a Design

How do you know when to use an independent-groups, repeated-measures, or matched-pairs design? There are times when it is obvious that a repeated-measures design is not appropriate. If the effects of one or both treatments are irreversible, as in comparing the effects of one type of brain surgery with another, counterbalancing is not possible. And, as we have already mentioned, when the carry-over effects are not equal in both directions, a repeated measures design cannot solve the problem. Often, however, using the same individuals in two conditions will be economical and useful; this will be true only when the threat of persisting carry-over effects can be eliminated.

An investigator will usually adopt a repeated-measures design in order to increase the sensitivity and power of his or her experiment. Some of the same power can be achieved by using a matched-pairs design, inasmuch as it also minimizes the influence of certain individual differences and, of course, completely overcomes the carry-over problem. But it is not always obvious what criteria to use for matching, and often matching requires extensive pretesting to obtain measures on the matching variables. In addition, when individuals are matched on one variable and assigned to conditions in pairs, there is still no assurance that other variables which might potentially influence their performance are distributed equally across the two conditions.

In general, then, when expected effects are small relative to the expected variability among individuals, a repeated-measures or matched-pairs design should be considered (but only used if it passes muster in terms of several limitations). When the expected effects are relatively large, an independent-groups design in which subjects are measured in only one condition to which they are assigned randomly is a safer procedure.

## Practice Problems

### *A. Oh, I Definitely Agree*

An investigator was interested in how the behavior of individuals is influenced by others. As part of a series of studies on conformity, the following experiments were conducted.

*Experiment 1.* On each trial, the subject's task was to judge which of three *comparison* lines was the same length as a *standard* line. All four lines were projected on a screen and were present while the subject made a choice. There were three other people in the room also making choices. Unknown to the subject, these three people had been previously instructed about what response to make on each trial (i.e., these students were confederates). There were 16 trials, and on a random half of these trials, the confederates made their choices before the subject, unanimously selecting a wrong comparison line. On the other half of the trials, the subject made his choice before the confederates responded. All judgments were made orally. After each trial, the experimenter recorded whether or not the subject had selected the line that actually matched the standard line in length. For each subject, then, the experimenter had the total number of correct responses on those trials when the subject responded first and on those trials when the subject responded last.

| Responded First | Responded Last |
|---|---|
| 5 | 3 |
| 6 | 2 |
| 7 | 4 |
| 8 | 3 |

Analyze these data to determine whether order of responding influenced the subject's performance.

*Experiment 2.* The subject's task was the same as in Experiment 1. There were 12 trials for each subject. Some subjects were tested alone, and other subjects were tested with three confederates present. In the treatment in which the confederates were present, they unanimously chose the wrong line, and the subject always responded last.

| Alone | Confederates Present |
|---|---|
| 5 | 1 |
| 8 | 1 |
| 5 | 4 |

Conduct an appropriate analysis of these data.

## *B. Which Way Is Up?*

The research division of an airplane company rank-ordered 16 pilots with respect to the number of hours of flying experience each had. One of the first two pilots was randomly assigned to Condition 1 and the other to Condition 2. Similarly, the third and fourth pilots were randomly assigned to Conditions 1 and 2, and so forth. The conditions corresponded to two possible designs of an aircraft instrument panel, and the score for each pilot was the number of errors on a simulated test flight.

| Condition 1 | Condition 2 |
|---|---|
| 8 | 7 |
| 6 | 6 |
| 1 | 3 |
| 5 | 7 |
| 3 | 6 |
| 2 | 1 |
| 7 | 5 |
| 4 | 4 |

What kind of a design is this? Did the type of instrument panel affect the pilots' performances?

## *C. And Now a Word From Our Sponsor?*

An investigator wanted to determine if external stimulation is one of the causes of dreaming. Each subject in the experiment spent two nights in the laboratory. On the first night, the subject slept under standard, Quiet conditions and on the second night, a tape recording of TV programs was turned on shortly after the subject fell asleep (the level was sufficiently low so that no person woke up). During both sessions, the investigator recorded the total number of minutes spent dreaming (assume this can be done by monitoring various physiological indices such as eye movements). Since the mean number of minutes dreaming was significantly greater during the Noisy Session than during the Quiet Session, the investigator concluded that external stimulation increases dreaming. Would you accept the conclusion? Explain.

## *D. The Numbers Game*

Design an experiment to look at the effects of incentives (e.g., monetary rewards) on problem solving, using (1) a random, independent-groups design, (2) a matched-pairs design, and (3) a repeated-measures design. Briefly describe the independent and dependent variables, subject-selection and -assignment procedures, and treatment conditions included in the study. Finally, speculate on the relative merits of the three methods of subject assignment in the context of this experiment.

# Variance Ratios:
## The $F$ Test

# 8

IN THIS CHAPTER, WE ARE GOING TO DEVELOP ANOTHER TEST for determining whether hypotheses of the form $\mu_1 = \mu_2$ can be rejected. The new test, referred to as the *F ratio*, or *F test*, is conceptually very similar to $t$ and can be understood readily in terms of concepts already familiar to you, such as sum of squares, variance, and degrees of freedom. One advantage of the $F$ test is that it can be used not only for cases involving two groups but also when three, four, or in fact any number of populations are being compared. For example, Table 8-1 lists the hypothetical data from an experiment designed to determine whether water source influences the growth rate of bean plants. The experimental *factor* of interest here was water source, and there were four *levels* of the experimental factor: tap, bottled, rain, and pond water. Four bean seeds were randomly assigned to each of the treatment conditions. The amount of water, type of soil, lighting, temperature, and other conditions were, of course, the same in the four treatments. The dependent variable was the amount of growth to the nearest inch after six weeks. Thus, the null hypothesis we want to test ($H_0$) is: $\mu_T = \mu_B = \mu_R = \mu_P$.

## ACCOUNTING FOR THE VARIABILITY AMONG ALL SCORES

Intuitively, the reasoning behind the $F$ test begins by examining all the scores we have obtained in an investigation. Consider the scores in Table 8-1. Our first observation about them might be that they are not all the same; rather, they differ or vary among themselves. Why? Two types of answers suggest themselves. The variability might occur because of the differential influence of type of water our plants received, the *treatment effect.* But we can also see quite readily that even within a particular treatment the scores differ among themselves, presumably because of random fluctuations in the size and quality of the seeds and soil used. This random within-group variability alone would certainly produce some differences between the treatment means by chance, even if the water did not really have any differential effects. Our job therefore becomes one of determining whether observed treatment differences are greater than would be expected from random variation and uncontrolled factors. And this is most conveniently done by creating a ratio, the $F$ ratio, to compare the mean differences among the groups with the variability among the scores within each group. If (but only *if*) the treatment differences are unlikely to be explained merely by chance variability among the scores, we will be able to reject the null hypothesis and conclude that water source had a real effect. This same logic will later be used to reduce very complex problems to simple terms, but for the moment, we need only expand on the basic idea by pursuing our first example in detail.

TABLE 8-1 Amount of growth in six weeks (to nearest inch) of bean plants under four water conditions

| | *Tap* | *Bottled* | *Rain* | *Pond* |
|---|---|---|---|---|
| | 5 | 4 | 7 | 6 |
| | 7 | 7 | 8 | 4 |
| | 4 | 6 | 8 | 8 |
| | 6 | 7 | 6 | 3 |
| $\Sigma$ | 22 | 24 | 29 | 21 |
| $\bar{X}$ | 5.50 | 6.00 | 7.25 | 5.25 |

## TESTING A HYPOTHESIS ABOUT MEANS WITH THE RATIO OF TWO VARIANCES

Let us first consider the case in which $H_0$ is true. Look at the scores in Table 8-1 within any particular treatment (e.g., the Tap Water condition). No two scores are the same, and these variations probably reflect uncontrolled factors such as soil content and many others that we could not have kept perfectly constant. Such uncontrolled variability in the population is often called *random*, or *experimental, error.*

Now examine the differences *among the means* of the four conditions. Why should the means of these samples differ if it is true that $\mu_T = \mu_B = \mu_R = \mu_P$? We have said that the means of the samples from populations with equal means can differ because of sampling fluctuations, again as a result of the fact that not all the individual scores within the population are the same (random or experimental error). In other words, the variability among subjects treated alike (or the *within*-group variability) provides an estimate of random or experimental error; the variability among the different conditions (the *between*-groups variability) also provides an estimate of the same random error. If we actually had numerical values for these two estimates, we could form the ratio

$$\frac{\text{Between-groups variability}}{\text{Within-group variability}}$$

If no real treatment factors contributed to between-groups variability (i.e., when $H_0$ is true), then we would expect the value of this ratio to be close to 1.00. Now consider the case in which $H_0$ is false. The scores within any particular condition will still differ because of random error. But the means of the samples will differ both because of sampling fluctuations or random error *and* because of differences among the populations caused by *treatment effects.* In the present example, if the source of the water influences growth, then there ought to be differences among the means of the various conditions over and above the differences that would be expected on the basis of chance fluctuations in sample means. Hence, the value of the ratio

$$\frac{\text{Between-groups variability}}{\text{Within-group variability}} = \frac{\text{random error} + \text{treatment effects}}{\text{random error}}$$

should be greater than 1.00 when $H_0$ is false. Thus, when the null hypothesis is false, the only thing that has changed is that there is an additional component contributing to the differences among means, namely, the treatment effect. This information is summarized in Table 8-2.

## SAMPLING DISTRIBUTION OF THE VARIANCE RATIO

Even when the null hypothesis is true, our ratio would not always be exactly equal to 1.00. We would expect some fluctuation in the value of this ratio simply as a consequence of chance. But at the same time, very large values of the ratio are quite unlikely. As you might have guessed, a conventional

TABLE 8-2 Summary of logic behind the $F$ test

| State of the World | |
|---|---|
| $H_0$ *True* | $H_0$ *False* |
| Variability among treatment means estimates random error | Variability among treatment means estimates random error *plus* real differences in populations or treatment effects |
| Variability among subjects treated alike estimates random error | Variability among subjects treated alike estimates random error |
| Expected ratio = 1.00 | Expected ratio > 1.00 |

decision rule is to reject the null hypothesis when the obtained value of this ratio is so large that it is unlikely to occur by chance more than 5 times in 100.

What we need to know, then, is the form of the sampling distribution of between-groups variability/within-group variability when $H_0$ is true. If you had plenty of time (and perhaps the help of a computer), you could find this empirically for any particular investigation in a fashion entirely analogous to that described for obtaining an empirical distribution of the difference between two means. First, set up four very large identical populations. Next, draw one random sample consisting of four cases from each population, and use these scores to compute the value of the ratio of between-groups variability/within-group variability. (The actual computational procedure for this will be described later). Next, take the value of this ratio, and plot it on a graph on which the horizontal axis represents possible values of the ratio and the vertical axis represents frequency of occurrence. Now, start all over again, drawing four new samples of four cases each, compute the ratio, and plot it. If you did this for a large number of such experiments when, in fact, the means of the four populations were equal, the sampling distribution you would obtain would look approximately like that in Figure 8-1a. With a little reflection, you will see that the general form of the distribution makes sense. Values close to 1.00 should be most probable. Values greater than 1.00 (which would lead us to suspect that a treatment effect was present) become increasingly less probable. A value in the portion of the distribution representing the 5 percent largest values would, by convention, lead us to reject the null hypothesis. (A value less than 1.00 would not be of any interest because the ratio can fall below 1.00 only as a result of random fluctuations in our sample and never as a result of the presence of treatment effects.)

Figure 8-1b shows similar sampling distributions for cases in which $\mu_1 = \mu_2 = \mu_3$ and the number in each sample is 4 and for cases in which $\mu_1 = \mu_2 = \mu_3 = \mu_4 = \mu_5 = \mu_6$ and the number in each sample is 4. Figure

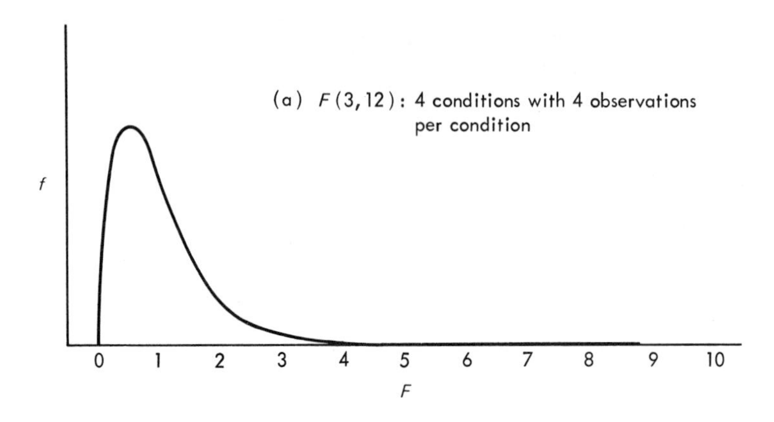


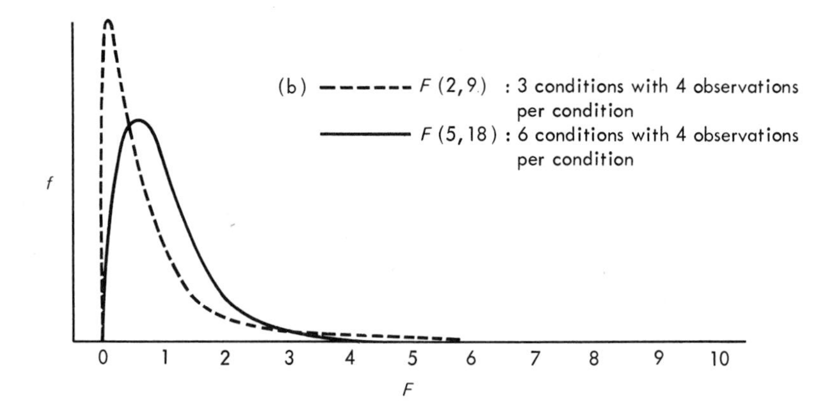


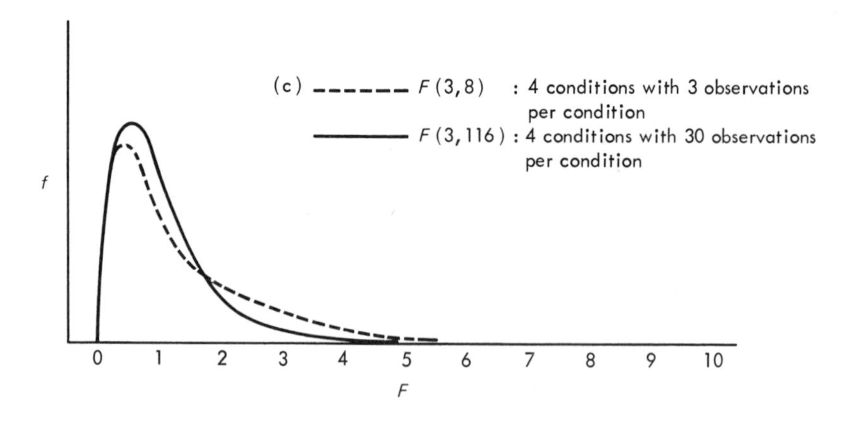


FIG. 8–1 Illustrative $F$ distributions: (a) 4 conditions, 4 observations per condition, (b) 3 or 6 conditions with 4 observations per condition, (c) 4 conditions with 3 or 30 observations per condition.

8-1c shows sampling distributions when $\mu_1 = \mu_2 = \mu_3 = \mu_4$ for cases in which the number in each sample is 3 and in which the number in each sample is 30. As you can see from Figure 8-1, the form of the ratio of between-groups variability/within-group variability depends upon the number of populations compared and upon the number of cases in each sample. In general, any particular large value of the ratio is less likely to occur by chance as the number of cases sampled increases and as the number of populations compared increases.

The family of theoretical distributions called $F$ distributions closely approximates the relative probabilities of ratios of various sizes and can thus be used as a model of what to expect in cases such as the one we have here. (The $F$ ratio is named for Sir Ronald Fisher, who developed the practical applications of this family of distributions.) As in the case of $t$ distributions, any particular $F$ distribution is specified in terms of its *degrees of freedom*. Each estimate of the random error has its own degrees of freedom; the numerator, or between-groups variability, has $df =$ number of groups $-1$. The denominator, or within-group variability, has

$$df = (n_1 - 1) + (n_2 - 1) + (n_3 - 1) + (n_4 - 1) + \cdots$$

and so on, up to the number of groups you have. Thus, the ratio of between- to within-groups variability in our plant experiment should fall into the $F$ distribution with 3 and 12 $df$, usually abbreviated $F\,(3, 12)$, when $\mu_1 = \mu_2 = \mu_3 = \mu_4$. The table in Appendix E gives certain critical values for a number of the members of the family of $F$ distributions. A value of $F$ greater than 3.49 is only likely to happen 5 times in 100 by chance in the case of $F\,(3, 12)$ when the null hypothesis is true. Does our obtained value of $F$ (where $F =$ between-groups variability/within-group variability) exceed 3.49?

## COMPUTING *F*

Remember, $F$ is the ratio of two estimates of variance; and to estimate a variance, we begin by computing an appropriate sum of squares ($SS$).

### Components of the Total Sum of Squares

We could get a sum of squares ($SS$) for all the scores we have, ignoring the condition from which they came, which would express the total variability among all our observations. To do this, each score is subtracted from the grand mean ($\bar{T}$) of all the scores; each of these deviations is squared; the squared deviations for all subjects are added together

$$SS_{\text{Total}} = \sum (X - \bar{T})^2$$

The deviation of any particular subject (e.g., subject *i* in Group 1) from the grand mean of all the scores can also be expressed as the sum of two components: the extent to which the subject deviates from its group mean $(X_i - \bar{X}_1)$ and the extent to which its group mean deviates from the grand mean $(\bar{X}_1 - \bar{T})$. Thus

$$X_i - \bar{T} = (X_i - \bar{X}_1) + (\bar{X}_1 - \bar{T})$$

This is shown graphically in Figure 8-2.

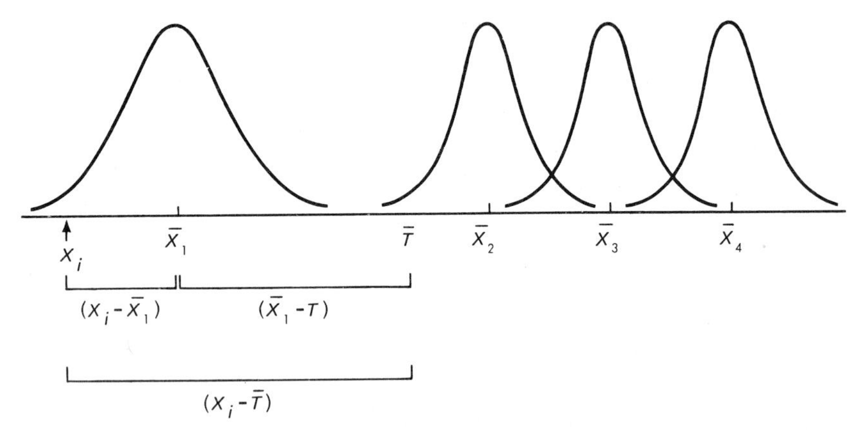


FIG. 8–2 Graphic representation showing that the deviation of each subject from the grand mean of all conditions consists of two parts: (1) the deviation of the subject from the mean of his group $(X_i - \bar{X}_1)$ and (2) the deviation of his group mean from the grand mean $(X_1 - \bar{T})$.

These deviation scores can be squared for each subject

$$(X_i - \bar{T})^2 = (X_i - \bar{X}_1)^2 + (\bar{X}_1 - \bar{T})^2$$

and added together for all subjects

$$\sum (X - \bar{T})^2 = \sum (X - \bar{X})^2 - \sum (\bar{X} - \bar{T})^2$$

Notice that the left-hand side of this equation is the $SS_{\text{Total}}$ and the right-hand side consists of two components, within-group *SS* and between-groups *SS*.

*Within-group SS*. The first component on the right side of the equation, $\sum (X - \bar{X})^2$, is called the *sum of squares within* $(SS_{\text{Within}})$. It reflects the extent to which individuals within a condition vary among themselves.

*Between-groups SS*. The second component on the right side of the equation is called the sum of squares between ($SS_{\text{Between}}$). It reflects the extent to which the means of the samples vary among themselves.*

These two components of $SS_{\text{Total}}$ provide the basis for the two estimates of the random error or uncontrolled variability that we need to compute the $F$ ratio.

*Computational formulas for the sums of squares.* The computation of these three sums of squares is simplified by using the following formulas:

$$SS_{\text{Between}} = \frac{\sum(\sum X)^2}{n} - \frac{(\sum X_T)^2}{N}$$

$$SS_{\text{Within}} = \sum X^2 - \frac{\sum(\sum X)^2}{n}$$

$$SS_{\text{Total}} = \sum X^2 - \frac{(\sum X_T)^2}{N}$$

(*Note*: $n$ = number of observations per condition, and $N$ = total number of observations.)

*Between-groups variance* (*mean square between*). The estimate of the variance in the population based on the deviations among means is not equivalent to the raw $SS$ because we must take into account the number of observations on which it is based. Specifically, the between-groups $SS$ is divided by the $df$ associated with it, namely, the number of treatments minus 1. This value is called the *mean square between*. Remember, the mean square between is a variance.

$$MS_{\text{Between}} = \frac{SS_{\text{Between}}}{df_{\text{Between}}} = \frac{SS_{\text{Between}}}{\text{number of treatments} - 1}$$

*Within-group variance* (*mean square within*). The estimate of the variance in the population based on the deviations among subjects treated alike is found by dividing the within-group $SS$ by the $df$ associated with it, namely, the number of subjects in each group minus 1, summed across all groups. This value is called the *mean square within*.

$$MS_{\text{Within}} = \frac{SS_{\text{Within}}}{df_{\text{Within}}} = \frac{SS_{\text{Within}}}{(n_1 - 1) + (n_2 - 1) + \cdots \text{etc.}}$$

This is essentially a pooled estimate of the population variance comparable

*The $SS_{\text{Between}}$ should really be called the $SS$ "among" whenever there are more than two means involved. However, we will follow tradition and always refer to the sum of squares between groups.

to that computed in calculating $t$ values ($s_p^2$). Thus, in our example $MS_{\text{Within}}$ could be obtained as follows:

$$= \frac{SS_1 + SS_2 + SS_3 + SS_4}{n_1 + n_2 + n_3 + n_4 - 4}$$

*F ratio.* The formula for the $F$ ratio is as follows:

$$F = \frac{MS_{\text{Between}}}{MS_{\text{Within}}}$$

The quantity $MS_{\text{Between}}/MS_{\text{Within}}$ is distributed as $F(df_{\text{Between}}, df_{\text{Within}})$.

## UNDERSTANDING THE SUMMARY TABLE

The information given in the preceding section is summarized in Table 8-3. Usually, the results of the corresponding calculations are also represented in such a summary table, as shown in Example 1.

TABLE 8-3 Summary of computations for one-factor independent-groups analysis of variance

| *Source* | *SS* | *df* | *MS* | *F* |
|---|---|---|---|---|
| Between | $\frac{\sum(\sum X)^2}{n} - \frac{(\sum X_T)^2}{N}$ | number of levels − 1 | $\frac{SS_{\text{Between}}}{df_{\text{Between}}}$ | $\frac{MS_{\text{Between}}}{MS_{\text{Within}}}$ |
| Within | $\sum X^2 - \frac{\sum(\sum X)^2}{n}$ | $(n_1 - 1) + (n_2 - 1) \ldots$ etc. | $\frac{SS_{\text{Within}}}{df_{\text{Within}}}$ | |
| Total | $\sum X^2 - \frac{(\sum X_T)^2}{N}$ | $N - 1$ | | |

Note: When the $n$'s are equal in each condition, the first term in the $SS_{\text{Between}}$ is

$$\Sigma\,(\Sigma\,X)^2/n = [(\Sigma\,X_1)^2 + (\Sigma\,X_2)^2 \ldots]/n.$$

When the $n$'s are not equal, the first term in the $SS_{\text{Between}}$ is

$$(\Sigma\,X_1)^2/n_1 + (\Sigma\,X_2)^2/n_2 + \ldots \text{etc.}$$

EXAMPLE 1

Now, let us actually analyze the data from our hypothetical experiment on the effects of water source on growth of bean plants. The data from Table 8-1 are given again in Table 8-4a, along with certain preliminary values that are necessary for computing the $F$ ratio. The actual computations are shown in Table 8-4b. It is good to get in the habit of performing analyses in a systematic and uniform fashion. If you always try to set up your data in the same way, analyses will go faster and will be easier to check for errors. This becomes more important as the complexity of the analysis increases.

TABLE 8-4 Hypothetical data and summary table for Example 1

(a)

| | *Tap* | *Bottled* | *Rain* | *Pond* | $\sum$ | $(\sum X)^2$ |
|---|---|---|---|---|---|---|
| | 5 | 4 | 7 | 6 | | |
| | 7 | 7 | 8 | 4 | | |
| | 4 | 6 | 8 | 8 | | |
| | 6 | 7 | 6 | 3 | | |
| $\bar{X}$ | 5.50 | 6.00 | 7.25 | 5.25 | | |
| $\sum X$ | 22 | 24 | 29 | 21 | 96 | 9216 |
| $\sum X^2$ | 126 | 150 | 213 | 125 | 614 | |
| $(\sum X)^2$ | 484 | 576 | 841 | 441 | 2342 | |

(b)

| *Source* | | *SS* | *df* | *MS* | *F* |
|---|---|---|---|---|---|
| Between | $\frac{2342}{4} - \frac{9216}{16} = 585.5 - 576 =$ | 9.5 | 3 | 3.17 | 1.33 |
| Within | $614 - 585.50 =$ | 28.5 | 12 | 2.38 | |
| Total | $614 - 576 =$ | 38.0✓ | 15✓ | | |

Because $1.33 < 3.49$ and $p > 0.05$,
we may not reject $H_0$: $\mu_T = \mu_B = \mu_R = \mu_P$

To make sure you are reading Tables 8-3 and 8-4 properly, we will work through each step in this example.

Between-Groups *SS*

$$SS_{\text{Between}} = \frac{\sum(\sum X)^2}{n} - \frac{(\sum X_T)^2}{N}$$

$$= \frac{22^2 + 24^2 + 29^2 + 21^2}{4} - \frac{96^2}{16}$$

$$= \frac{484 + 576 + 841 + 441}{4} - \frac{9{,}216}{16}$$

$$= \frac{2{,}342}{4} - \frac{9{,}216}{16}$$

$$= 585.5 - 576$$

$$= 9.5$$

Notice that $\sum(\sum X)^2$, 2,342, and $(\sum X)^2$, 9,216, can be taken directly from the work sheet. Also notice that each numerator is divided by the number of scores which were summed *prior* to the squaring operation in the numerator. In the first term, 4 scores added up to 22, which was squared; 4 scores added up to 24, which was squared; and so forth. Similarly, in the second term, 16 scores added up to 96, which was squared.

### Within-Groups *SS*

$$
\begin{aligned}
SS_{\text{Within}} &= \sum X^2 - \frac{\sum(\sum X)^2}{n} \\
&= \frac{5^2 + 7^2 + 4^2 + \cdots + 4^2 + 8^2 + 3^2}{1} - \frac{22^2 + 24^2 + 29^2 + 21^2}{4} \\
&= \frac{126 + 150 + 213 + 125}{1} - \frac{2{,}342}{4} \\
&= \frac{614}{1} - \frac{2{,}342}{4} \\
&= 614 - 585.5 \\
&= 28.5
\end{aligned}
$$

Notice that 614 is taken from the work sheet and is simply the sum of each squared score. Also notice that 585.5 is the value of the *first term* in $SS_{\text{Between}}$.

### Total *SS*

$$
\begin{aligned}
SS_{\text{Total}} &= \sum X^2 - \frac{(\sum X)^2}{N} \\
&= \frac{5^2 + 7^2 + 4^2 + \cdots + 4^2 + 8^2 + 3^2}{1} - \frac{96^2}{16} \\
&= \frac{126 + 150 + 213 + 125}{1} - \frac{9{,}216}{16} \\
&= 614 - 576 \\
&= 38.0
\end{aligned}
$$

Notice that 576 is the value of the *second term* in the $SS_{\text{Between}}$. The check mark (√) next to 38.0 in Table 8-4b indicates that we have verified that

$$
\begin{gathered}
SS_{\text{Total}} = SS_{\text{Between}} + SS_{\text{Within}} \\
38.0 = 9.5 + 28.5
\end{gathered}
$$

If this relationship does not hold (excluding minor rounding error), then there is an error in the calculations, and they should be done over.

### *df*

Notice that $df_{\text{Total}} = N - 1$ and that

$$df_{\text{Total}} = df_{\text{Between}} + df_{\text{Within}}$$
$$15 = 3 + 12$$

If the degrees of freedom have been properly assigned, they will add up to the total number of scores minus 1.

### Variance Estimates (Mean Squares)

$$MS_{\text{Between}} = \frac{SS_{\text{Between}}}{df_{\text{Between}}} = \frac{9.5}{3} = 3.17$$

$$MS_{\text{Within}} = \frac{SS_{\text{Within}}}{df_{\text{Within}}} = \frac{28.5}{12} = 2.38$$

### *F* Ratio

Under the null hypothesis, $MS_{\text{Between}}$ and $MS_{\text{Within}}$ are two estimates of random error (population variances). The ratio of these two estimates is called $F$ because the ratio is distributed as the theoretical $F$ distribution with $(df_{MS\text{Between}}, df_{MS\text{Within}})$.

$$F = \frac{MS_{\text{Between}}}{MS_{\text{Within}}} = \frac{3.17}{2.38} = 1.33$$

### Decision

If $\mu_T = \mu_B = \mu_R = \mu_P$, then $F$ (3, 12) will exceed the value of 3.49 by chance 5 percent of the time. Because our obtained $F$, 1.33, is less than 3.49, we may not reject $H_0$ at the .05 level. There is no evidence that the type of water affects growth rate of bean plants.

EXAMPLE 2

Table 8-5a gives data from a hypothetical study in which samples from three populations of kindergarten children were compared: (1) children who

TABLE 8-5 Data and summary table for Example 2

(a)

| | *No Day Care* | *Free Play* | *Structured Activities* | | |
|---|---|---|---|---|---|
| | 0 | 1 | 2 | | |
| | 1 | 2 | 2 | | |
| | 0 | 0 | 3 | | |
| | 2 | 3 | 3 | | |
| | 3 | 2 | 4 | | |
| | 1 | 2 | 3 | | |
| $\bar{X}$ | 1.17 | 1.67 | 2.83 | | |
| | | | | $\Sigma$ | $(\Sigma X)^2$ |
| $\Sigma X$ | 7 | 10 | 17 | 34 | 1,156 |
| $\Sigma X^2$ | 15 | 22 | 51 | 88 | |
| $\Sigma (X)^2$ | 49 | 100 | 289 | 438 | |

(b)

| *Source* | | *SS* | *df* | *MS* | *F* |
|---|---|---|---|---|---|
| Between | $\frac{438}{6} - \frac{1,156}{18} = 73 - 64.2 =$ | 8.8 | 2 | 4.4 | 4.4 |
| Within | $88 - 73 =$ | $= 15$ | 15 | 1.0 | |
| Total | $88 - 64.2 =$ | $= 23.8$✓ | 17✓ | | |

Because $4.4 > 3.68$ and $p < .05$, we can reject $H_0$: $\mu_N = \mu_{FP} = \mu_{SA}$.

had never been to a day-care center; (2) children who had attended a day-care center for at least six months, during which most of the time was spent in free-play activities; and (3) children who had attended a day-care center for at least six months, during which most of the time was spent in structured activities. The measure of interest was the number of cooperative acts (e.g., helping other children, sharing, cleaning up) exhibited by each child during a three-day observation period. The children's behavior was rated by observers who were blind with respect to the pre-experimental history of the children.

The obtained mean numbers of cooperative acts were 1.17, 1.67, and 2.83 for No–Day Care, Free-Play, and Structured-Activities conditions, respectively. Is the evidence sufficient to reject the hypothesis that $\mu_N = \mu_{FP} = \mu_{SA}$? An appropriate analysis is indicated in Table 8-5b. Because the critical value of $F(2, 15) = 3.68$ at the .05 level (according to Appendix E) and our obtained

$F$ is 4.40, we may reject the null hypothesis. The amount of cooperation shown is not equal in all three populations of children. Remember, we did not assign the children to the three groups at random; therefore, we may not conclude that day-care training influences cooperativeness. For example, it is possible that parents who are most concerned about raising cooperative children also are more likely to send their children to day-care centers with structured activities.

EXAMPLE 3

Analyze the data in Table 5-3 (the study on the relationship of performance in the first-semester medical school classes and anxiety level) by computing an $F$ ratio. Verify the values summarized in Table 8-6. You will always arrive

TABLE 8-6 Example from Table 5-1 as an *F* test

| *Source* | *SS* | *df* | *MS* | *F* |
|---|---|---|---|---|
| Between | 579 − 541.5 = 37.5 | 1 | 37.50 | 8.17 |
| Within | 680 − 579 = 101.0 | 22 | 4.59 | |
| Total | 680 − 541.5 = 138.5 | 23 | | |

at the same decision (within the limits of rounding errors) whether you analyze a two-group investigation by computing $t$ or by computing $F$. This is because the two analyses use basically the same information from the data. In fact, you could think of $t$ as similar to the ratio of between-groups differences to within-group differences. There is also an invariant relationship between the theoretical $t$ and $F$ distributions such that $t^2 = F$, when two samples are being compared (i.e., when $df_{\text{Between}} = 1$).

## Practice Problems

### A. *Candy, Tea, or Me?*

A study was conducted to compare the effectiveness of different rewards that might be used in teaching retarded children. Twenty retarded children, ages five to seven, were randomly assigned to four independent groups. Each child was shown five common objects and five cards, each showing the printed name of one of the objects. The child's task was to match each object correctly with its name card. Whenever a correct match was made, the experimenter rewarded the child. Children in the first group were rewarded with candy; children in the second group were rewarded with tokens that could later be exchanged for candy; children in the third group were rewarded with tokens that could later be exchanged for attention from the experimenter (playing games, reading to the child, and so on); children in the fourth group were rewarded with verbal praise. The experimenter recorded the number of trials required before a child could correctly match all five pairs. The scores were:

| Candy | Tokens for Candy | Tokens for Attention | Praise |
|---|---|---|---|
| 9 | 4 | 6 | 11 |
| 7 | 5 | 3 | 7 |
| 6 | 6 | 5 | 7 |
| 7 | 5 | 4 | 5 |
| 6 | 5 | 4 | 6 |

Analyze these data. Be sure to set up an appropriate summary table. Did the type of reward affect the children's performance?

## *B. A Is Taller Than B. C Is Taller Than A. Who's Higher, C or B?*

In order to resolve a controversy over which type of coffee (Peoria Gold, Biloxi Beaut, or Salinas Smooth) produces the greatest stimulating effect, eleven students conducted a study. One person was randomly selected to serve as the experimenter, and the other ten were randomly assigned to three beverage conditions. Each subject drank one cup of coffee in the late afternoon (only the experimenter knew which treatment a subject received) and then worked on logic problems for thirty minutes. The number of correct solutions for each subject is given below. Note: the $n$'s are not equal (see bottom of Table 8-3).

| Peoria Gold | Biloxi Beaut | Salinas Smooth |
|---|---|---|
| 5 | 2 | 3 |
| 3 | 4 | 4 |
| 3 | 1 | 2 |
| | 3 | |

Does the evidence indicate that the type of coffee influences performance on logic problems? If you could add a condition to this study, what would it be?

## *C. The Numbers Game*

Verify that for Practice Problem A in this set, $H_0$ can be rejected if $\alpha = .05$, but not if $\alpha = .01$. Explain what this means. Because we would like to be as certain as possible of the correctness of our inferences, why don't we always set $\alpha = .01$, or better yet, .001?

# Comparisons Among Treatment Means in Single-Factor Experiments

# 9

THE BASIC CONCEPTS UNDERLYING $F$ TESTS, introduced in Chapter 8, provide the foundation for many useful statistical analyses appropriate to a wide variety of research designs. In Chapter 8, we used an $F$ ratio specifically to test the null hypothesis that the population means of several groups were equal. When we use $F$ in this way and find a significant value, we can reject hypotheses of the form, $H_0$: $\mu_1 = \mu_2 = \mu_3 = \ldots = \mu_n$. Such a finding does not, however, help us locate the source of our significant effect. We know that overall something happened, but we do not know exactly what. For example, recall the experiment investigating the effects of water source on the growth rate of bean plants. Below are two sets of hypothetical means for such an experiment:

| | *Tap* | *Bottled* | *Rain* | *Pond* |
|---|---|---|---|---|
| Set 1 | 5.50 | 5.25 | 8.30 | 5.36 |
| Set 2 | 5.50 | 5.25 | 8.30 | 8.36 |

Both of these sets might result in a significant overall $F$ value. Something happened in each case, but not the same something; the two sets of data clearly show different patterns. In Set 1, there does not appear to be much difference among tap, bottled, and pond water, but rainwater seems superior to the other three conditions. In Set 2, there does not appear to be much difference between tap and bottled water or between rain and pond water, but the latter two conditions seem to result in a higher growth rate than the former two do.

When more than two groups are involved, the $F$ obtained from a simple one-factor analysis of variance does not differentiate the various patterns of outcomes. In such cases, follow-up statistical analyses are often needed. This chapter discusses these procedures. They are really variations of what you already know, but they allow you to ask much more specific questions of the data than simply whether or not all the means are equal.

## COMPARISONS OF INDIVIDUAL GROUPS

First, let us consider an experiment involving three conditions designed to answer two very specific questions. Suppose you wanted to evaluate a new cold remedy. You might randomly assign people coming into a health center complaining of colds to two conditions: those who are told simply to go home and rest and those who are told to rest *and* to take Un-stuff. Let us assume that you ask all the people to come back to the health center the next day and that your measure or dependent variable is each person's rating of how well he or she feels after one day of your remedy. However, this might not be your complete design. After receiving an unfamilar or novel treatment, people will often feel better simply because they think they should or because they think they are being helped and thus actually do feel better. Improvements of this sort, which cannot be attributed specifically to the primary aspect of the treatment in which you are interested (in this case, consequences of the pharmacological compound Un-stuff), are called *placebo effects*. In order to determine whether there are any placebo effects operating in your experiment, you might include a third condition in which a random selection of people with colds are told to go home and rest and to take a remedy that looks like Un-stuff in every way but that, unknown to them, is a chemically inert substance such as a sugar pill. Your complete design would now include three conditions: no-treatment control, placebo control, and Un-stuff treatment. In addition, before you collect your data, you would have in mind two specific questions that, in effect, determined your design: (1) Is there a placebo effect in this situation? (2) Is there any treatment effect above and beyond this potential placebo effect? The first question would be answered by comparing the scores of the No-Treatment people with those of the Placebo-Control subjects, in other words, by testing the hypothesis $H_0$: $\mu_{NTC} - \mu_{PC} = 0$. The second question would be answered by comparing the

Placebo-Control condition with the Un-stuff condition ($H_0$: $\mu_{PC} - \mu_U = 0$). Both comparisons can be made by forming an $F$ ratio based on a sum of squares for the specific groups being compared and the within-group sum of squares for the entire experiment. The data for this hypothetical experiment are given in Table 9-1, along with the relevant preliminary calculations.

TABLE 9-1 Data from an experiment on the effects of a remedy for colds

| | *No Treatment* | *Placebo* | *Un-stuff* | |
|---|---|---|---|---|
| | 10 | 7 | 7 | |
| | 8 | 7 | 6 | |
| | 7 | 6 | 5 | |
| | 8 | 7 | 5 | |
| | 9 | 8 | 6 | |
| | 9 | 6 | 4 | |
| | 7 | 4 | 5 | |
| | 8 | 5 | 6 | |
| | 7 | 5 | 3 | |
| | 7 | 5 | 3 | |
| $\bar{X}$ | 8.00 | 6.00 | 5.00 | |
| $\sum X$ | 80 | 60 | 50 | 190 |
| $\sum X^2$ | 650 | 374 | 266 | 1290 |
| $(\sum X)^2$ | 6400 | 3600 | 2500 | 12500 |

First, the within-group sum of squares (which, remember, is a pooled estimate of the variability of subjects treated alike and which is therefore an estimate of random error) is found in the usual way (see Table 8-3, page 112).

$$SS_{\text{Within}} = \sum X^2 - \frac{\sum (\sum X)^2}{n}$$

$$= 10^2 + 8^2 + \cdots + 3^2 + 3^2 - \frac{80^2 + 60^2 + 50^2}{n}$$

$$= 1290 - \frac{12500}{10}$$

$$= 1290 - 1250$$

$$= 40$$

Next, we need to find the sum of squares for each comparison of interest. Each is essentially a between-groups sum of squares for the two conditions of the experiment that are relevant to the comparison. Remember, the general form of the computational formula for a between-groups $SS$ (see Table 8-3, page 112) is

$$SS_{\text{Between}} = \frac{\sum (\sum X)^2}{n} - \frac{(\sum X_T)^2}{N}$$

For Comparison 1, no treatment versus placebo control,

$$SS_{\text{comp 1}} = \frac{80^2 + 60^2}{10} - \frac{(80 + 60)^2}{20}$$

Note especially that the $\sum X_T$ equals the sum of the totals of the two conditions entering into the comparison (80 + 60) and that $N$ equals the sum of the number of observations in each of the two conditions entering into the comparison (10 + 10 = 20).

$$\begin{aligned} SS_{\text{comp 1}} &= \frac{6400 + 3600}{10} - \frac{140^2}{20} \\ &= 1000 - 980 \\ &= 20 \end{aligned}$$

Similarly, the $SS$ for the second comparison, contrasting the Placebo-Control condition with the Un-stuff condition, is

$$\begin{aligned} SS_{\text{comp 2}} &= \frac{60^2 + 50^2}{10} - \frac{(60 + 50)^2}{20} \\ &= \frac{3600 + 2500}{10} - \frac{110^2}{20} \\ &= 610 - 605 \\ &= 5 \end{aligned}$$

The information we have thus far ($SS_{\text{comp 1}}$, $SS_{\text{comp 2}}$, and $SS_{\text{Within}}$) is listed in the first column of Table 9-2.

TABLE 9-2 Individual comparisons on the data from Table 9-1

| | $SS$ | $df$ | $MS$ | $F$ |
|---|---|---|---|---|
| Comparison 1 (no treatment versus placebo control) | 20 | 1 | 20 | 13.51 |
| Comparison 2 (placebo control versus un-stuff) | 5 | 1 | 5 | 3.38 |
| Within | 40 | 27 | 1.48 | — |

The second column lists the *df* associated with each of these sums of squares. Because each comparison includes two groups, the *df* equals number of conditions minus 1 (2 − 1 = 1). The *df* for the within-group *SS* is based on information from all three conditions:

$$(n_1 - 1) + (n_2 - 1) + (n_3 - 1) = (10 - 1) + (10 - 1) + (10 - 1) = 27.$$

The mean squares are given in the third column and are found in the usual way, by dividing each *SS* by its *df*.

The *F* ratio for each comparison (given in the fourth column) is found by dividing the mean square associated with the comparison by the $MS_{\text{Within}}$.

$$F_{\text{comp 1}} = \frac{20.00}{1.48} = 13.51$$

$$F_{\text{comp 2}} = \frac{5.00}{1.48} = 3.38$$

According to Appendix E, the critical value of *F* at the .05 level with 1 and 27 *df* is 4.21. Because $13.51 > 4.21$, we may reject $H_0\colon \mu_{NT} - \mu_{PC} = 0$; because $3.38 < 4.21$, we may not reject $H_0\colon \mu_{PC} - \mu_U = 0$. Therefore, we may conclude that our experiment indicated that when people took a pill, they felt better, but we may not conclude that the real medication produced any significant effect beyond that of the placebo. Un-stuff just did not work.

## COMPARISONS WITH COMBINED GROUPS

Other types of comparisons are frequently of interest to researchers. For example, sometimes you might want to compare the performance of one condition with the average performance of two or more other conditions. Imagine that you were interested in the effects of early experiences on the learning ability of young rats. Assume that you randomly assigned 8 animals to each of three conditions. In the Standard condition, the animals were kept in individual cages with only a simple exercise wheel. In the Enriched Environment 1 condition, the animals were placed in individual cages that contained not only an exercise wheel but also other toys that the animal could push, pull, and carry. In the Enriched Environment 2 condition, the animals were placed in individual cages with an activity wheel and a structure of platforms, tunnels, and ladders that provided opportunities for climbing and exploring. After three months in their respective cages, each rat was presented with a maze-learning problem, and you recorded the number of errors made during ten training sessions. The data are listed in Table 9-3.

Prior to conducting this study, you probably would have had two primary questions in mind: (1) Does raising rats in enriched environments have any consequences for their performance on a learning task? (2) Is one of the enriched environments more beneficial than the other? The first question can be translated into a statistical hypothesis in the following form:

$$\text{Comparison 1 } H_0\colon \mu_S - \left(\frac{\mu_{E1} + \mu_{E2}}{2}\right) = 0$$

That is, is the average performance of the Enriched 1 and Enriched 2 conditions equal to that of the Standard condition? The second question is of the

TABLE 9-3 Data from an experiment on the effects of type of living environment on learning

| | Number of Errors | | | |
|---|---|---|---|---|
| | *S* | *E1* | *E2* | |
| | 7 | 5 | 3 | |
| | 6 | 5 | 5 | |
| | 5 | 6 | 4 | |
| | 6 | 5 | 3 | |
| | 6 | 4 | 5 | |
| | 5 | 5 | 3 | |
| | 6 | 6 | 4 | |
| | 7 | 4 | 5 | |
| $\bar{X}$ | 6.00 | 5.00 | 4.00 | |
| $\sum X$ | 48 | 40 | 32 | 120 |
| $\sum X^2$ | 292 | 204 | 134 | 630 |
| $(\sum X)^2$ | 2304 | 1600 | 1024 | |

type (discussed in the preceding section) in which the means of two groups are compared.

$$\text{Comparison 2 } H_0\colon \mu_{E1} - \mu_{E2} = 0$$

Again, we need a sum of squares for each comparison that, along with the within-group SS, will form the basis of an *F* test.

To answer the first question, we want to compare the Standard condition with the *combined* enriched conditions. Because $\sum X_{E1} = 40$ and $\sum X_{E2} = 32$, the combined $\sum X_{E1+E2} = 72$. Similarly, $n_{E1} = 8$ and $n_{E2} = 8$; therefore, $N_{E1+E2} = 16$. Thus, $\bar{X}_{E1+E2} = \frac{72}{16} = 4.50$. This is equivalent to $(\bar{X}_{E1} + \bar{X}_{E2})/2$ when the *n*'s are equal in the two conditions $[(5 + 4)/2 = 4.50]$.

Our $H_0$, remember, is $\mu_S - (\mu_{E1} + \mu_{E2})/2 = 0$. The statistical test still essentially involves comparing *two* samples: the Standard condition with the Enriched (Combined) condition. We will designate this latter group *EC*. Now, we follow the procedure for a comparison of two conditions.

$$\begin{aligned}
SS_{\text{comp 1}} &= \frac{\sum (\sum X)^2}{n} - \frac{(\sum X)^2}{N} \\
&= \left[\frac{(\sum X_S)^2}{n_S} + \frac{(\sum X_{EC})^2}{n_{EC}}\right] - \frac{(\sum X_T)^2}{N} \\
&= \left(\frac{48^2}{8} + \frac{72^2}{16}\right) - \frac{120^2}{24} \\
&= (288 + 324) - 600 \\
&= 612 - 600 \\
&= 12
\end{aligned}$$

The *SS* for the comparison of *E*1 with *E*2 is described on page 121–123.

$$SS_{\text{comp 2}} = \frac{\sum (\sum X)^2}{n} + \frac{(\sum X_T)^2}{N}$$

$$= \frac{40^2 + 32^2}{8} - \frac{72^2}{16}$$

$$= 328 - 324$$

$$= 4$$

The $SS_{\text{Within}}$ is straightforward and based on the information from all three conditions

$$SS_{\text{Within}} = \sum X^2 - \frac{\sum (\sum X)^2}{n}$$

$$= 630 - \frac{48^2 + 40^2 + 32^2}{8}$$

$$= 630 - 616$$

$$= 14$$

The $SS_{\text{comp 1}}$, $SS_{\text{comp 2}}$, and $SS_{\text{Within}}$ are given in the first column of Table 9-4. The remainder of the analysis is exactly the same as that described in the preceding section and is summarized in the remaining columns of Table 9-4.

TABLE 9-4 Analysis of data from Table 9-3, illustrating both combined and individual comparisons

| *Source* | *SS* | *df* | *MS* | *F* |
|---|---|---|---|---|
| Comparison 1 (*S* vs. *EC*) | 12 | 1 | 12 | 17.91 |
| Comparison 2 (*E*1 vs. *E*2) | 4 | 1 | 4 | 5.97 |
| Within | 14 | 21 | 0.67 | |

One possible point of confusion is the *df* associated with Comparison 1. You should note that because conceptually only *two* conditions are being compared, Standard and Enriched (Combined), the *df* is number of conditions minus 1, or 2 − 1 = 1.

According to Appendix E, the critical value of $F\ (1, 21) = 4.32$; both comparisons are therefore significant. Thus, the evidence from this experiment indicates that the enriched environments facilitated learning and that *E*2 was superior to *E*1.

## CONTROLLING THE ERROR RATE FOR MULTIPLE COMPARISONS

When we set an alpha level such as .05 for a one-factor analysis of variance (described in Chapter 8), we ensure that when the null hypothesis is true we will falsely reject $H_0$ no more than 5 percent of the time. If we did only experiments that were analyzed in this fashion, our probability of a Type I error (given $H_0$ is true) would be .05 *for each experiment*. But whenever we perform more than one statistical test on a set of data, as in the case of the comparisons outlined in the two preceding sections, our error rate *per experiment* may increase.

Suppose, for example, that you performed an experiment on new instructional methods with three groups of kindergarten children, that two of the groups each received a different new method you were testing, and that the third group served as a control and received only normal kindergarten instruction. Suppose, further, that (unknown to you) neither of the new methods improves learning but that, by chance, the children in the Control group were less bright than those in the Treatment groups.* If you now compare the first Treatment group with the Control group, you will obtain a significant difference, say at $p < .05$, because of the unusually poor performance of the Control group rather than because the new treatment worked. There is no way you could know this, however, and so you would make a Type I error. The error could not be avoided, but you may take comfort in noting that such an error will occur less than 5 times in 100.

But now a hitch arises. You will almost certainly want to compare the second Treatment group with the same Control group used in your first comparison, and these poor performers will once again make the treatment you are testing look good—better than it really is. So you now have made a second Type I error because you involved the same Control group in two comparisons. This is roughly how multiple comparisons can increase your real error rate. Under some circumstances, the error rate for an experiment increases even more as comparisons are added.

The fact that the possibility of committing a Type I error increases with the number of comparisons made in an experiment does not necessarily discourage investigators from conducting more complex experiments. For example, our test of the cold remedy could have been conducted as a two-group experiment in which only no treatment and Un-stuff were compared. Thus, only one statistical test would have been necessary, and our error rate for the experiment would have been .05. However, we saw that we needed to include a third group (placebo control) and that two comparisons were

*This could happen by chance even if you randomly assigned children to groups; in fact, the presence of occasional flukes is the reason we must use statistics and compute the probability of making errors in the first place.

necessary (no treatment versus placebo and placebo versus Un-stuff) in order to understand fully the type of effect that our cold remedy might have. Still, conducting two comparisons as opposed to one overall $F$ ratio approximately doubled our Type I error rate for this experiment.

There are at least two possible ways to handle this increase in error rate as the number of comparisons increases. (1) Figure that the increase in information from more complex experiments is worth the increased risk of a Type I error, and do not worry about it. (2) Use various techniques to *hold down* the error rate for an experiment when more than one test is made. For example, we could decide to set a more stringent alpha level for each comparison. If we required our obtained $F$ to fall in the most extreme .025 of the distribution for each of the two comparisons in the cold-remedy experiment, then the total probability of a Type I error for the whole experiment would have been closer to .05. In practice, both ways of dealing with potential increases in Type I error rates are used and which predominates depends largely on two considerations.

First, were the comparisons of interest generated *prior* to the collection of the data, and did they dictate the design of the experiment; or were they generated *after* the study was run and after data analysis began? If each comparison answers a clearly formed question that has been framed in advance, then many investigators think it is acceptable to tolerate the increase in potential error rate that results from several comparisons. However, if the questions arise later and lead to additional comparisons after the data have been examined, most investigators try to hold down the error rate in the experiment as these comparisons are added. One rationale for this distinction between *a priori* and *post hoc** comparisons is that differences which you expected and can justify on theoretical or logical grounds are less likely to be produced by chance than differences you did not expect.

Second, what is the number of comparisons being made? When many comparisons are conducted, and especially when you use the data from one or more conditions repeatedly, most investigators select a procedure for keeping the total Type I error rate for the experiment to some acceptable level. Although there are many techniques that have been developed to hold down the error rate for a study when multiple comparisons are made, we will present only two commonly used methods as examples.

### Comparisons Among All Treatment Means (Scheffé's Test)

Suppose you managed a business and wanted to compare the performance of 6 of your new salespeople. Five customers were randomly assigned to each salesperson, and the number of items sold was recorded. The data are given in Table 9-5a.

**Post hoc* comparisons are sometimes referred to as *a posteriori* comparisons.

TABLE 9-5 (a) Number of sales per customer from a hypothetical study of differences among salespeople

| | Salesperson | | | | | | |
|---|---|---|---|---|---|---|---|
| | *A* | *B* | *C* | *D* | *E* | *F* | |
| | 4 | 3 | 2 | 4 | 0 | 7 | |
| | 2 | 3 | 3 | 6 | 2 | 6 | |
| | 3 | 2 | 3 | 5 | 2 | 7 | |
| | 2 | 0 | 2 | 5 | 3 | 6 | |
| | 3 | 1 | 1 | 5 | 1 | 5 | |
| $\bar{X}$ | 2.80 | 1.80 | 2.20 | 5.00 | 1.60 | 6.20 | |
| $\sum X$ | 14 | 9 | 11 | 25 | 8 | 31 | 98 |
| $\sum X^2$ | 42 | 23 | 27 | 127 | 18 | 195 | 432 |
| $(\sum X)^2$ | 196 | 81 | 121 | 625 | 64 | 961 | 2,048 |

(b) analysis

| *Source* | *SS* | *df* | *MS* | *F* |
|---|---|---|---|---|
| Between | $\frac{2,048}{5} - \frac{98^2}{30}$ | | | |
| | $409.6 - 320.13 = 89.47$ | 5 | 17.89 | 19.24 |
| Within | $432 - 409.6 = 22.4$ | 24 | 0.93 | |
| Total | $432 - 320.13 = 111.87$✓ | 29✓ | | |

In this situation, you would not necessarily have had any hypotheses in advance about who is better than whom. Therefore, your first task would be to perform an analysis of variance to determine if there are any differences at all among the salespeople. The results of this analysis are given in Table 9-5b; as practice, you might do the analysis yourself and check your answer.

Because the overall $F$ indicates that some differences exist among conditions (salespeople), you would probably want to look at the data in more detail. You might want to compare each person with each other person. For example, A with B, A with C, A with D, A with E, and A with F; B with C, B with D, and so on. In addition, you might have noticed after the experiment that three of your salespeople are under forty (A, D, and E) and three are over forty years of age (B, C, and F). Thus, you might want to compare the *combined* under-forty data with the *combined* over-forty data. Similarly, if four of the people dressed casually (A, B, C, and E) and two more formally (D and F), you might generate a comparison of the combined casual versus the combined formal dressers.

The procedure for obtaining the $F$ ratio for any of these comparisons is the same as we have described previously. Of the many potential comparisons, the two suggested in the preceding paragraph (A, D, and E versus B, C, and

TABLE 9-6 Additional statistical analyses of data from Table 9-5

(*a*) *Comparison 1: Younger Versus Older* (*A*, *D*, *and E versus B*, *C*, *and F*)

$$SS = \frac{\sum (\sum X)^2}{n} - \frac{(\sum X_T)^2}{N}$$

$$= \frac{(14 + 25 + 8)^2 + (9 + 11 + 31)^2}{15} - \frac{98^2}{30}$$

$$= \frac{47^2 + 51^2}{15} - \frac{98^2}{30}$$

$$= \frac{2209 + 2601}{15} - \frac{98^2}{30}$$

$$= 320.67 - 320.13$$

$$= 0.54$$

(*b*) *Comparison 2: Casual Versus Formal* (*A*, *B*, *C*, *and E versus D and F*)

$$SS = \left[\frac{(\sum X_1)^2}{n_1} + \frac{(\sum X_2)^2}{n^2}\right] - \frac{(\sum X_T)^2}{N}$$

$$= \left[\frac{(14 + 9 + 11 + 8)^2}{20} + \frac{(25 + 31)^2}{10}\right] - \frac{98^2}{30}$$

$$= \left(\frac{42^2}{20} + \frac{56^2}{10}\right) - \frac{98^2}{30}$$

$$= (88.2 + 313.6) - 320.13$$

$$= 401.8 - 320.13$$

$$= 81.67$$

(*c*) *Summary Table*

| *Source* | *SS* | *df* | *MS* | *F* |
|---|---|---|---|---|
| Comparison 1 | 0.54 | 1 | 0.54 | 0.58 |
| Comparison 2 | 81.67 | 1 | 81.67 | 87.82 |
| Within | 22.40 | 24 | 0.93 | |

F and A, B, C, and E versus D and F) are illustrated in Tables 9-6a and 9-6b, respectively.

Once each of these $F$ ratios has been found, it would ordinarily be compared with the critical value of $F$ (1, 24) = 4.26. However, because these comparisons were generated after the fact and used the data from some groups repeatedly, we must control the Type I error rate for the experiment. As we have already mentioned, one way to do this is to set a more stringent alpha level for each comparison. This is equivalent to requiring a larger value of the $F$ ratio for each comparison in order to reject $H_0$. In fact, the *corrected* critical value of $F$ ($F_s$) is given by the following formula:

$$F_s = [F(df_{\text{Between}}, df_{\text{Within}})](a - 1)$$

where $F(df_{\text{Between}}, df_{\text{Within}})$ = critical value of $F$ for the overall analysis (shown in Table 9-5b) and $a$ = number of conditions in the study. In the present example

$$F_s = 2.62\,(6 - 1)$$
$$= 13.10$$

Thus, Comparison 2 is significant ($87.82 > 13.10$), but Comparison 1 is not ($.58 < 13.10$).

This procedure is called the *Scheffé test* after the person who developed it (hence, the subscript $s$ is used for this $F$), and it sets a very stringent criterion for rejecting any particular $H_0$. This is because the $F$ value required for rejection is so large that all possible comparisons between two groups or between combined conditions could be made and the total probability of a Type I error for the experiment would not exceed, in our case, 5 percent. Others have worked out less conservative (i.e., less stringent) tests that are especially appropriate for particular situations, and they all have the common property of reducing the alpha level for each comparison when more than one comparison is made. The next section presents an example of a procedure that serves the same general purpose, but in the special situation in which many different groups are individually compared with a single standard.

### Comparing Each of Several Conditions with a Control or Standard Condition (Dunnett's Test)

A special test (*Dunnett's test*) is available for the case in which an investigator wants to compare each of several conditions with a standard or control condition. For example, suppose you were investigating the brain areas involved in the regulation of water intake in white mice. Assume you were trying to show the importance of Area A by demonstrating that stimulation of this area (with a mild electric current through implanted electrodes) increases the amount of water drunk by animals as compared with the amount drunk by animals who were not stimulated. In order to rule out the possibility that electrical stimulation of *any* area of the brain increases drinking, you might also include two additional groups of animals: stimulation of Area B and stimulation of Area C. The data for this hypothetical experiment are given in Table 9-7a.

The first step in analyzing these data would be to conduct an overall analysis of variance. A summary of this analysis is given in Table 9-7b. As additional practice, you might work through this analysis and check your answer. Because the critical value of $F(3, 24) = 3.01$ at the .05 level and the obtained value of $F = 10.50$, we may reject the hypothesis $H_0\colon \mu_{NS} = \mu_A = \mu_B = \mu_C$.

Next, we would probably want to compare each of the stimulation conditions with the No-Stimulation condition to determine whether stimulation

TABLE 9-7 Data and analysis for a hypothetical experiment dealing with the effects of brain stimulation on water intake of white mice

(a) Number of Units of Water Drunk in Test Period

| | *Area Stimulated* | | | | |
|---|---|---|---|---|---|
| | *NS* | *A* | *B* | *C* | |
| | 5 | 8 | 6 | 4 | |
| | 3 | 9 | 4 | 5 | |
| | 4 | 5 | 6 | 3 | |
| | 2 | 7 | 4 | 3 | |
| | 4 | 6 | 5 | 6 | |
| | 6 | 10 | 6 | 2 | |
| | 5 | 8 | 4 | 1 | |
| $\bar{X}$ | 4.14 | 7.57 | 5.00 | 3.43 | |
| $\sum X$ | 29 | 53 | 35 | 24 | 141 |
| $\sum X^2$ | 131 | 419 | 181 | 100 | 831 |
| $(\sum X)^2$ | 841 | 2809 | 1225 | 576 | 5451 |

(b) Analysis

| *Source* | *SS* | *df* | *MS* | *F* |
|---|---|---|---|---|
| Between | $\frac{5{,}451}{7} - \frac{141^2}{28}$ <br> $778.71 - 710.04 = 68.67$ | 3 | 22.89 | 10.50 |
| Within | $831 - 778.71 = 52.29$ | 24 | 2.18 | |
| Total | $831 - 710.04 = 120.96$✓ | 27✓ | | |

of each area resulted in a significant change in amount of water consumed. This is where Dunnett's test is useful. To begin, compute an $F$ ratio for each comparison of interest (Comparison 1 is $H_0$: $\mu_{NS} - \mu_A = 0$; Comparison 2 is $H_0$: $\mu_{NS} - \mu_B = 0$; Comparison 3 is $H_0$: $\mu_{NS} - \mu_C = 0$) in the usual way. The calculations are summarized in Table 9-8.

Next, we apply Dunnett's correction to hold down the value of $\alpha$ for the experiment. Because the Dunnett tables are based on critical values of $t$ rather than $F$, we must convert our $F$ ratios to $t$. Remember (page 117) $t^2 = F$, when two samples are being compared. Therefore

$$t_{\text{comp 1}} = \sqrt{18.88} = 4.34$$
$$t_{\text{comp 2}} = \sqrt{1.18} = 1.09$$
$$t_{\text{comp 3}} = \sqrt{0.82} = 0.90$$

Dunnett has worked out a special table that indicates the value of $t$ needed to reject each null hypothesis in the set. This value depends upon the $df$ for the $MS_{\text{Within}}$ and the number of conditions in the experiment. As you can see in Appendix $F$, for any given $df$ associated with the $MS_{\text{Within}}$, the critical

TABLE 9-8 Additional statistical analyses of data from Table 9-7

(a) Comparison 1 (*NS* versus *A*).

$$SS = \frac{29^2 + 53^2}{7} - \frac{82^2}{14}$$
$$= 521.43 - 480.28$$
$$= 41.15$$

(b) Comparison 2 (*NS* versus *B*).

$$SS = \frac{29^2 + 35^2}{7} - \frac{64^2}{14}$$
$$= 295.14 - 292.57$$
$$= 2.57$$

(c) Comparison 3 (*NS* versus *C*).

$$SS = \frac{29^2 + 24^2}{7} - \frac{53^2}{14}$$
$$= 202.43 - 200.64$$
$$= 1.79$$

(d) Summary

| *Source* | *SS* | *df* | *MS* | *F* |
|---|---|---|---|---|
| Comparison 1 (*NS* vs. *A*) | 41.15 | 1 | 41.15 | 18.88 |
| Comparison 2 (*NS* vs. *B*) | 2.57 | 1 | 2.57 | 1.18 |
| Comparison 3 (*NS* vs. *C*) | 1.79 | 1 | 1.79 | 0.82 |
| Within | 52.29 | 24 | 2.18 | |

value of $t$ increases as the number of conditions (and thus comparisons with the standard group) increases. With only two conditions (the control and one experimental), the critical value of $t$ is exactly the same as it is for an ordinary $t$ test. You can verify this by comparing the first column in Appendix F with Appendix D.

In our example, there are four treatment conditions and 24 *df* for the $MS_{\text{Within}}$; the critical value of $t$ for each of our comparisons is 2.51. Therefore, only Comparison 1 is significant, and we may conclude that only brain stimulation in Area A significantly affected the amount of water consumed by our animals.

## COMPARING CORRECTED AND UNCORRECTED CRITICAL VALUES OF *F*

Suppose we had tested the three comparisons in our water-intake example with a simple uncorrected comparison procedure. In that case, the critical value of $F$ (1, 24) = 4.26. With Dunnett's correction, the critical value of $F = (2.51)^2 = 6.30$ (because according to Dunnett's table the critical value of $t = 2.51$ and $F = t^2$). With Scheffé's correction, the critical value of $F$ is $F_s = [F(3, 24)](a - 1) = 3.01 \times 3 = 9.03$. As you can see, Dunnett's test is less stringent than Scheffé's test because a smaller value of $F$ is required for rejection at the .05 level. The less severe correction is justified on the grounds that only a subset of all possible comparisons is being tested (e.g., the subset consisting of each experimental group compared with the control group). Dunnett's table provides the correction that will hold $\alpha$ at a specified level (e.g., .05) for this limited set of comparisons, whereas Scheffé's correction fixes the alpha level at the specified value assuming that all possible comparisons among treatment conditions are made.

In practice, deciding whether to apply a correction when more than one comparison is made for a set of data and, if so, which correction to use is partly a matter of the judgment of an individual investigator. The primary issue is how much risk of a Type I error is acceptable. If you allow the overall alpha level for an experiment to be too large, you cannot really have much confidence (much less persuade others!) that all your significant comparisons reflect real experimental effects. However, if you set very stringent criteria for rejecting any particular null hypothesis, you run the risk of missing or failing to detect potentially interesting and important real experimental effects.

## Practice Problems

### A. *Time to Think*

A psychologist believed that memories are established partly as a consequence of a process of protein synthesis in the brain. Puromycin is a chemical that supposedly interferes with protein synthesis when injected into the brain. The following study was an attempt to determine if retention is related to the time available for protein synthesis after training. Fifteen goldfish learned to avoid shock by swimming across a barrier into a safe chamber when a warning signal was presented. After an interval of several days, the animals' retention of the avoidance response was tested. The experimental factor was time between the end of training and the injection of puromycin. The intervals were 1 minute, 30 minutes, 60 minutes, and 90 minutes. There was also a condition in which no puromycin was administered. Three fish were randomly assigned to each of the five conditions. The number of correct responses on the retention test are listed below.

| 1 minute | 30 minutes | 60 minutes | 90 minutes | No Puromycin |
|---|---|---|---|---|
| 1 | 1 | 2 | 6 | 6 |
| 2 | 0 | 0 | 4 | 5 |
| 1 | 2 | 2 | 5 | 5 |

Analyze these data, including whatever a priori and/or post hoc comparisons you think are needed. What do these results suggest?

## *B. Visions of Sugar Plums*

The following data were obtained from a study of memory of nine-year-olds. Each child saw a sequence of either 12 words representing common objects or 12 pictures corresponding to the words. Furthermore, the pictures were either all black-and-white line drawings, black-and-white photographs, or color photographs. The children were assigned randomly to the four conditions, and the number of items each child could remember is listed below.

| Words | Line drawings | Black-and-white photographs | Color photographs |
|---|---|---|---|
| 6 | 8 | 7 | 10 |
| 5 | 7 | 7 | 9 |
| 5 | 6 | 8 | 10 |
| 4 | 6 | 6 | 7 |
| 6 | 8 | 5 | 8 |
| 3 | 9 | 8 | 9 |

Conduct appropriate analyses of these data, and describe the results.

## *C. The Numbers Game*

Look again at the data from the study of the relationship of type of day-care activities to cooperative behavior. (Table 8-5, page 116). Follow up the significant overall $F$ (see Table 8-5b) with (1) Dunnett's test and (2) Scheffé's test. Explain why you could come to different conclusions about this study with these two tests. Which do you think is more appropriate in this case?

# Factorial Designs: Basic Concepts

# 10

THE EXAMPLES WE HAVE PRESENTED SO FAR have been limited to studies of a single variable or factor (e.g., level of illumination, type of day-care activity, type of water). However, there are probably few events in the real world that depend only on a single variable. For example, the amount of cooperative behavior shown by a child will probably be determined not only by the type of day-care center the child has attended but also by the type of home environment in which the child has been raised, the youngster's age and sex, and numerous other factors. Similarly, the time it will take to hypnotize a person will probably depend not only on the level of illumination in the room but also on what the hypnotist says to the person and the person's prior experience with hypnosis. In addition to the type of water, potential variables that might affect the growth rate of bean plants include the amount of water, the type of soil, the hours of light per day, and numerous other things. What is more, the influence of any one variable may depend upon a second; the optimum amount of water for growing beans may depend upon which particular soil composition is used.

Fortunately, research designs and appropriate statistical analyses have been developed that allow us to study the simultaneous and combined effects of two or more variables or factors. The information obtained from one of these more complex designs is actually much greater than that yielded by one (or even several) simple, single-factor experiments.

Suppose you were interested in the factors that influence first-year college students' attitudes toward living in dormitories. In a simple two-group experiment investigating the effects of the size of the room, you might randomly assign some new students to small rooms and some new students to large rooms. At the end of the first semester, you could have each student rate the desirability of living on campus. Sample data for this hypothetical experiment are given in Table 10-1. The group means are $\bar{X}_S = 3.50$ and $\bar{X}_L = 7.50$. On the basis of a *t* test,* we could reject $H_0: \mu_S - \mu_L = 0$ and thus conclude that room size affected the ratings of the students, with larger rooms leading to higher ratings.

TABLE 10-1 Data from a one-factor experimental study of the effects of room size on college students' attitudes toward dormitory life

| | Room Size | |
|---|---|---|
| | *Small* | *Large* |
| | 6 | 7 |
| | 5 | 8 |
| | 6 | 8 |
| | 7 | 9 |
| | 2 | 8 |
| | 1 | 7 |
| | 1 | 6 |
| | 0 | 7 |
| $\Sigma X$ | 28 | 60 |
| $\bar{X}$ | 3.50 | 7.50 |

Suppose you did a second experiment, this time varying the number of roommates assigned to each person. Sample data comparing students with one and two roommates are given in Table 10-2. The means are $\bar{X}_1 = 7.00$ and $\bar{X}_2 = 4.00$, and the outcome of the *t* test indicates that we could reject $H_0: \mu_1 - \mu_2 = 0$ and conclude that the number of people sharing a room also affected the rated desirability of dorm life, with fewer roommates leading to higher ratings.

From these two simple experiments, we would know that larger rooms produce more positive ratings, as do fewer roommates. However, we would

*We could also perform an *F* test on these data.

TABLE 10-2 Hypothetical data from a one-factor experimental study of the effects of number of roommates on college students' attitudes toward dormitory life

| | Number of Roommates | |
|---|---|---|
| | *One* | *Two* |
| | 6 | 2 |
| | 5 | 1 |
| | 6 | 1 |
| | 7 | 0 |
| | 7 | 8 |
| | 8 | 7 |
| | 8 | 6 |
| | 9 | 7 |
| $\sum X$ | 56 | 32 |
| $\bar{X}$ | 7.00 | 4.00 |

know nothing about the *joint action* of these two factors. Will increasing the size of the rooms increase ratings above and beyond what might be obtained by reducing the number of people assigned to a room? Does the effect of increasing the number of roommates depend upon the size of the room?

We can begin to answer questions such as these with a research design called a *factorial design*, in which room size and number of roommates can be studied simultaneously. In the simplest factorial design, two levels of room size (e.g., small and large) and two levels of number of roommates (e.g., one and two) would be combined to yield the four treatment conditions indicated in the four cells of Table 10-3: (1) students living in small rooms with one roommate, (2) students living in small rooms with two roommates, (3) students living in large rooms with one roommate, (4) students living in large rooms with two roommates. In a random independent-groups design, each person in our sample would be assigned randomly to one of these four conditions. A sample set of raw scores is given in Table 10-4a. The outcome of the experiment can be more readily seen in Table 10-4b, which presents the means for each treatment condition within the cells of the table.

There are three major questions that can be asked with this design: (1) Is there an effect of the first factor (Factor A); in this case, is there an effect of room size? (2) Is there an effect of the second factor (Factor B); in this case, is there an effect of number of roommates? (3) Is there an interaction between Factor A and Factor B; for example, does the effect of increasing the number of roommates depend upon whether students are living in large or small rooms? Conversely, does the effect of increasing the size of the room depend upon the number of students assigned to it?

TABLE 10-3 The four groups of an experiment dealing with the effects of room size and number of roommates on desirability of dormitory life, arranged as a factorial design

| *Number of Roommates* | Room Size | |
|---|---|---|
| | *Small* | *Large* |
| One | Small/one | Large/one |
| Two | Small/two | Large/two |

TABLE 10-4 Data for the experiment shown in Table 10-3

(a) Raw-score Desirability Ratings

| | *Small/One* | *Small/Two* | *Large/One* | *Large/Two* |
|---|---|---|---|---|
| | 6 | 2 | 7 | 8 |
| | 5 | 1 | 8 | 7 |
| | 6 | 1 | 8 | 6 |
| | 7 | 0 | 9 | 7 |
| $\sum X$ | 24 | 4 | 32 | 28 |
| $\bar{X}$ | 6.00 | 1.00 | 8.00 | 7.00 |

(b) Mean Desirability Ratings

| *Number of Roommates* | Room Size | | |
|---|---|---|---|
| | *Small* | *Large* | |
| One | 6.00 | 8.00 | 7.00 |
| Two | 1.00 | 7.00 | 4.00 |
| | 3.50 | 7.50 | |

## MAIN EFFECTS

The first two questions raised in the preceding paragraph correspond to the questions asked in our simple two-group experiments (Tables 10-1 and 10-2). First, let us assess the effect of Factor A (room size) independent of Factor B (number of roommates) within the context of our factorial design. In order to determine whether the size of the room has an effect, we want to compare the mean ratings of all students living in large rooms with all students living in small ones. These values (given at the bottom of Table 10-4b) are $\bar{X}_S = 3.50$ and $\bar{X}_L = 7.50$. The value of $\bar{X}_S$ is obtained by adding up all the scores in the Small/One and Small/Two conditions and dividing by 8, or because the *n*'s in the conditions are equal

$$\bar{X}_S = \frac{\bar{X}_{S/1} + \bar{X}_{S/2}}{2} = \frac{6.00 + 1.00}{2} = 3.50$$

The value of $\bar{X}_L$ is found in the same way.

$$\bar{X}_L = \frac{\bar{X}_{L/1} + \bar{X}_{L/2}}{2} = \frac{8.00 + 7.00}{2} = 7.50$$

A statistical comparison of these two values will tell us whether there was a main effect of room size. The term *main effect* is used to refer to an effect of one variable evaluated without regard to the other variable (i.e., by summing, or *collapsing*, across levels of the second variable).

Our statistical question is whether $\bar{X}_S - \bar{X}_L = 3.50 - 7.50 = -4.00$ is sufficiently different from 0 to allow us to reject $H_0: \mu_S - \mu_L = 0$. The statistical analysis for evaluating the obtained difference is basically the same procedure you are familiar with from Chapter 9. (You may find it helpful to think of the present procedure as a comparison between two combined groups.) The actual computational procedure for the analysis for a two-factor design will be presented in a later section. For purposes of the present discussion, we will assume that all our obtained differences between means are statistically significant. Thus, overall, there was a main effect of room size, with larger rooms producing higher ratings.

The main effect of Factor B (number of roommates) is evaluated in a similar fashion. The mean of the combined One-Roommate conditions

$$\bar{X}_1 = \frac{\bar{X}_{S/1} + \bar{X}_{L/1}}{2} = \frac{6.00 + 8.00}{2} = 7.00$$

is compared with the mean of the combined Two-Roommates conditions

$$\bar{X}_2 = \bar{X}_{S/2} + \bar{X}_{L/2} = \frac{1.00 + 7.00}{2} = 4.00$$

These values are given on the right-hand side of Table 10-4b. Thus, there is a main effect of Factor B, indicating that increasing the number of roommates decreased ratings.

One advantage of the factorial design is that it is more economical than two separate experiments. In order to have eight observations per treatment, each of our two simple experiments would have required sixteen participants, for a total of thirty-two. The same information was obtained from our factorial design with only sixteen people.

## INTERACTIONS

In addition to economy, there is an even more important advantage of the factorial design. This is the unique opportunity to determine whether two (or more) variables interact. Two factors interact when the effect of one depends upon the specific level of the other factor.

In order to determine whether there is an interaction in our example, we will evaluate the effect of increasing room size *separately at each level* of the Number-of-Roommates variable. For students living with one other person, the mean difference in ratings between large and small rooms was

$$\bar{X}_{L/1} - \bar{X}_{S/1} = 8.00 - 6.00 = +2.00.$$

In contrast, the effect of increasing room size for students living with two other people was to increase the average rating by 6.00 points.

$$\bar{X}_{L/2} - \bar{X}_{S/2} = 7.00 - 1.00 = +6.00.$$

Thus, the effect of increasing the size of the room was not the same ($2.00 \neq 6.00$) at both levels of the Number-of-Roommates variable; hence, the two factors interact. In general, *two variables interact whenever the effect of one variable is not the same at every level of the other variable.*

Comparing the presence of an interaction with its absence will help to clarify this concept. Table 10-5 gives a new sample set of data for our dormitory example. Again, there are main effects of both room size

$$\bar{X}_L - \bar{X}_S = 5.50 - 3.50 = +2.00$$

and number of roommates

$$\bar{X}_2 - \bar{X}_1 = 2.00 - 7.00 = -5.00$$

However, the two variables do not interact. This is because the effect of increasing the size of the room for students with one roommate ($\bar{X}_{L/1} - \bar{X}_{S/1} = 8.00 - 6.00 = 2.00$) is the same as the effect of increasing the size of the room for students with two roommates ($\bar{X}_{L/2} - \bar{X}_{S/2} = 3.00 - 1.00 = 2.00$). Thus, when two variables do not interact, any particular statement about either of them based on the outcome of the analyses of main effects (e.g., ratings under Small-Room conditions are 2 units lower than those under Large-Room conditions) will be generally applicable. This statement is

TABLE 10-5 Alternative data for the hypothetical experiment shown in Table 10-3

(a)

| | *Small/One* | *Small/Two* | *Large/One* | *Large/Two* |
|---|---|---|---|---|
| | 6 | 2 | 7 | 4 |
| | 5 | 1 | 8 | 3 |
| | 6 | 1 | 8 | 2 |
| | 7 | 0 | 9 | 3 |
| $\sum X$ | 24 | 4 | 32 | 12 |
| $\bar{X}$ | 6.00 | 1.00 | 8.00 | 3.00 |

(b)

| | Room Size | | |
|---|---|---|---|
| *Number of Roommates* | *Small* | *Large* | |
| One | 6.00 | 8.00 | 7.00 |
| Two | 1.00 | 3.00 | 2.00 |
| | 3.50 | 5.50 | |

equally true for both levels of the Number-of-Roommates variable. However, when an interaction is present, any statement about the effect of one variable must be qualified by information about the other variable.

### Graphical Representations of Factorial Outcomes

It is often easier to see the outcome of a complex study when the data are presented graphically. Our two examples are plotted in Figure 10-1; the format is one frequently used for factorial designs. The dependent variable is always represented on the vertical axis. Either factor may be assigned to the horizontal axis. The second factor is represented by lines in the body of the graph. Each point corresponds to a treatment condition.

In the two graphs on the left of the figure, the main effect of room size in both experiments is indicated by the fact that the average of the two points on the right (above "Large") is higher than the average of the two points on the left (above "Small"). The general upward sweep of the lines also results from this main effect. The main effect of number of roommates reflects the fact that the average of the two points connected by the line labeled "One" is higher than the average of the two points connected by the line marked "Two." Also notice that the line connecting the One conditions is higher than the line connecting the Two conditions.

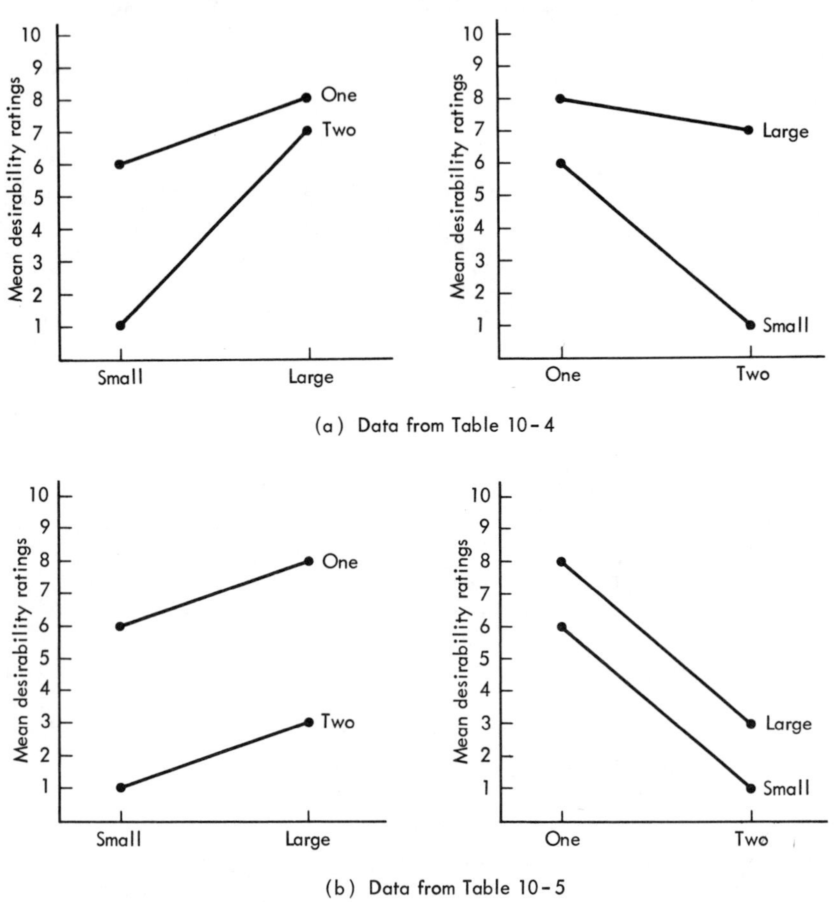


FIG. 10–1 Graphic representation of the means shown in Tables 10-4 and 10-5. Note that either factor can be placed on the horizontal axis.

The interaction in the first set of data (Figure 10-1a) is indicated graphically by the fact that the lines are not parallel. The slope of the line labeled "One" reflects the effect of varying room size for students assigned one roommate, and the slope of the line labeled "Two" reflects the effect of varying room size for students assigned two roommates. Because these slopes are not equal, the effects of varying room size are not equal at each level of the Number-of-Roommates factor. Conversely, whenever the slopes of the lines are equal (and the lines therefore are parallel), as in the second set of data (Figure 10-1b), the two factors do not interact with each other.

The graphs on the right of Figure 10-1 are simply alternative ways of plotting the same data, this time with number of roommates assigned to the horizontal axis.

## IMPORTANT USES OF FACTORIAL DESIGNS

Frequently, variables are combined factorially in order to determine the generality of the effects of one of the variables. An important subset of such studies is a situation in which an experimentally manipulated treatment variable is combined factorially with a *subject* variable (e.g., age, sex, IQ, type of difficulties experienced) in order to determine the generality over different populations of an experimental effect. For example, you might want to compare two types of therapy: group and individual. People with various problems could be assigned at random to the two therapy conditions, and the number of therapy sessions necessary to help each individual could be recorded. The outcome of this two-group study would tell you whether, on the average, one therapy was superior to the other. However, more information might be obtained if you built into your study a subject variable that would allow you to assess any potential differences in the relative effectiveness of the two therapies across types of problems. Imagine that you randomly selected participants from two clinical populations: people who are afraid to fly in airplanes and people who are depressed. Half of each group would then be assigned randomly to each type of therapy, yielding the design outlined in Table 10-6. A hypothetical set of means for this study is given in the cells of Table 10-6, and the data are graphed in Figure 10-2.

TABLE 10-6 Data from a hypothetical factorial experiment investigating the effect of types of therapy for two clinical problems

| | Type of Therapy | | |
|---|---|---|---|
| *Type of Problem* | *Group* | *Individual* | |
| Fear of Flying | 22.00 | 12.00 | 17.00 |
| Depression | 16.00 | 16.00 | 16.00 |
| | 19.00 | 14.00 | |

As you can see by inspecting the table and the figure, individual therapy, overall, required fewer sessions to achieve satisfactory results than group therapy did ($\bar{X}_I = 14.00$; $\bar{X}_G = 19.00$); and in general, those persons who were initially depressed were helped more quickly than those who were afraid of flying ($\bar{X}_D = 16.00$; $\bar{X}_F = 17.00$). However, the presence of an interaction indicates that these main effects do not give a complete picture in this case. The relative effectiveness of group and individual therapy depends upon the type of problem. Whereas individual therapy was superior to group

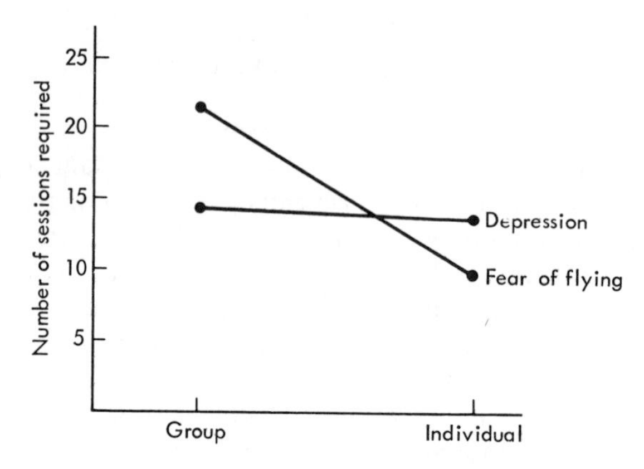


FIG. 10–2 Graphic representation of the means shown in Table 10-6.

therapy by 10 sessions with people who were afraid of flying, there was no difference between therapies in their effectiveness in treating people who were depressed. The interaction between type of therapy and type of problem in this case indicates that the superiority of individual to group therapy may be limited to people from specific clinical populations and, hence, is not a general phenomenon.

Sometimes, factorial designs are critical to the hypothesis an investigator is trying to test, and some other type of design simply cannot answer the question. Suppose, for example, that an educator found a number of children whose reading skills were below what would be expected from their IQ scores and other measures of ability. The investigator then suspects that these children have a perceptual disorder which prevents them from filtering out irrelevant information. To test this hypothesis, 20 normal children and 20 children from the problem group were selected randomly and asked to perform a task requiring perceptual ability. Specifically, each child was asked to count the number of dots appearing in a display window on a number of successive trials. The number of dots varied randomly from trial to trial, and the average time it took each child to count the dots was recorded. The dots were brightly colored, and no two dots were the same color on any given trial. If the perceptual hypothesis is correct, children from the problem group should be more distracted by the color (which was irrelevant to the task of counting) than the normal children and therefore should count more slowly. Suppose, too, $\bar{X}_N = 2.30$ seconds, and $\bar{X}_P = 6.23$ seconds. Can we conclude that our hypothesis is supported?

Upon careful analysis, it becomes apparent that our argument is not complete. The children in the Problem group might simply count more slowly than the children in the Normal group, irrespective of the presence of potentially distracting irrelevant colors. To rule out this alternative hypothesis, we could have had half of each group of children count colored dots and half count dots that did not vary in color (e.g., were all black). The design and three potential outcomes of this new, expanded factorial study are given in Figure 10-3.

The interaction depicted in Outcome A is one that would support our hypothesis. The problem children are no slower in the Black-Dot condition, but they are slower than normal children in the Colored-Dot condition. The interaction depicted in Outcome B would also support our hypothesis. Although the children from the Problem group count both types of dots more slowly, they are *especially* slow on the colored dots. Both sets of results are consistent with the hypothesis that the children from the Problem group have a harder time ignoring irrelevant features of the stimulus (in this case,

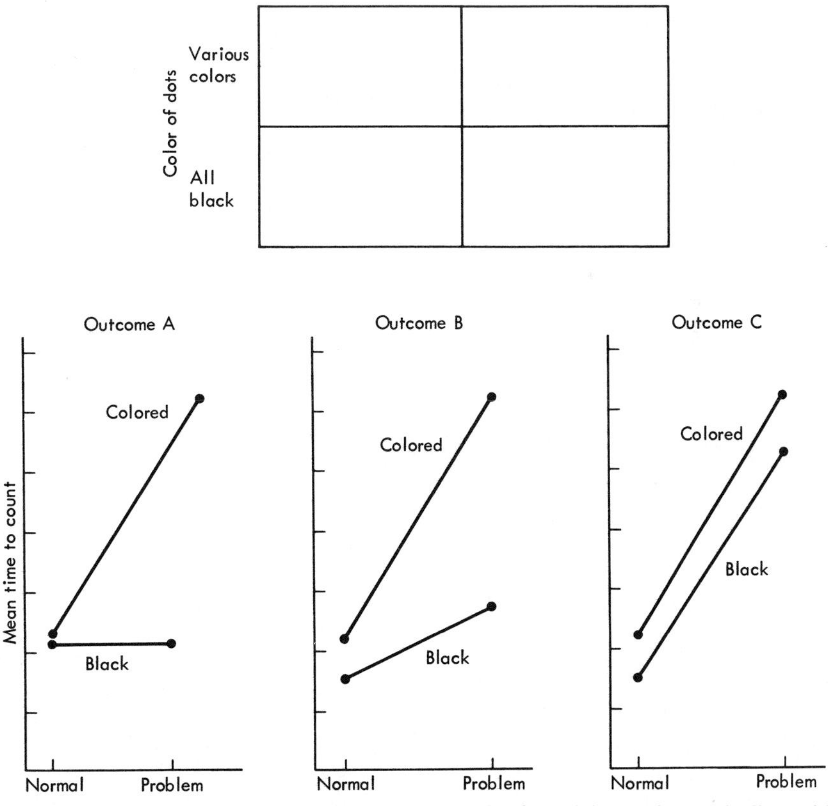


FIG. 10–3 Design and three possible outcomes of a factorial experiment dealing with the perceptual abilities of normal and problem children.

color) than normal children do. However, Outcome C would not support our initial hypothesis because both types of children were equally disrupted by the introduction of colored dots.

## ADDITIONAL PRACTICE FINDING MAIN EFFECTS AND INTERACTIONS

Figure 10-4 gives some potential outcomes for a two-factor experiment with two levels ($a_1$ and $a_2$) of Factor A and two levels ($b_1$ and $b_2$) of Factor B. Look at the table of means and the graphic representation of each set of

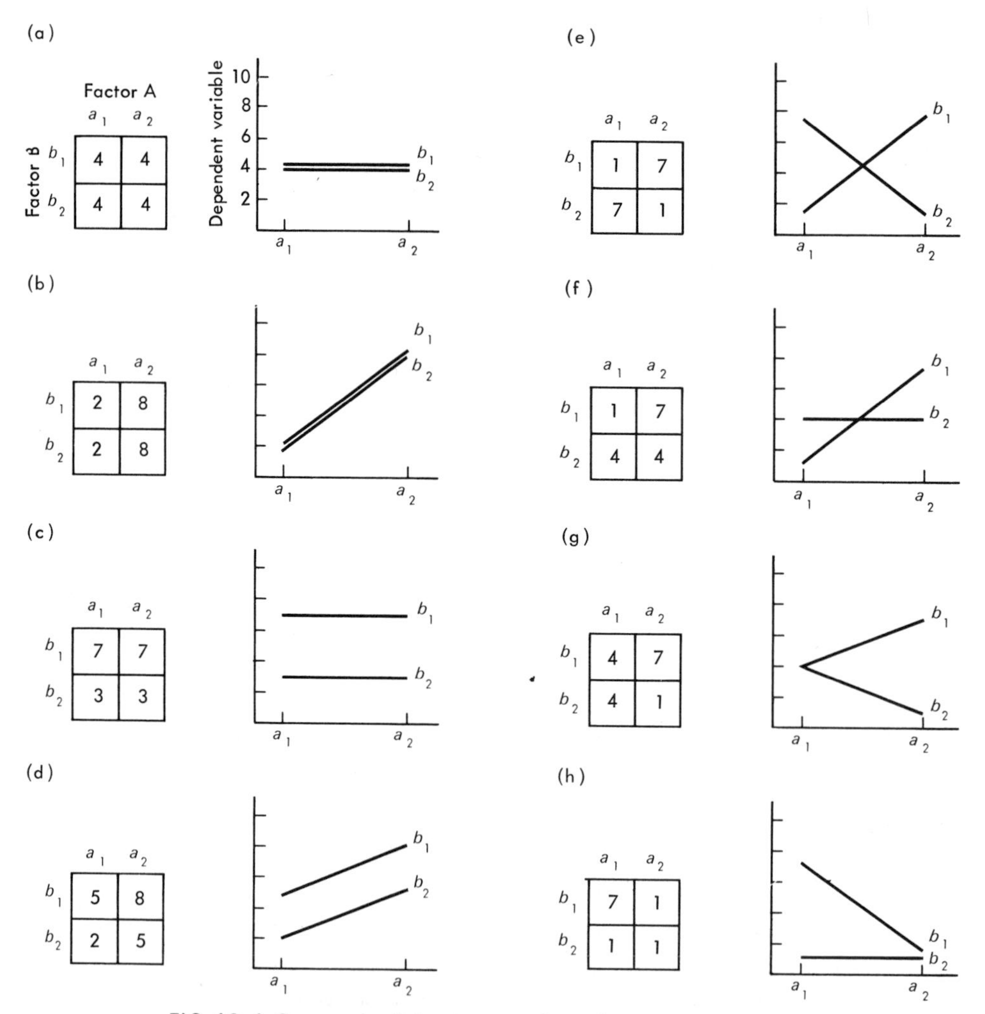


(a)

| | $a_1$ | $a_2$ |
|---|---|---|
| $b_1$ | 4 | 4 |
| $b_2$ | 4 | 4 |

(b)

| | $a_1$ | $a_2$ |
|---|---|---|
| $b_1$ | 2 | 8 |
| $b_2$ | 2 | 8 |

(c)

| | $a_1$ | $a_2$ |
|---|---|---|
| $b_1$ | 7 | 7 |
| $b_2$ | 3 | 3 |

(d)

| | $a_1$ | $a_2$ |
|---|---|---|
| $b_1$ | 5 | 8 |
| $b_2$ | 2 | 5 |

(e)

| | $a_1$ | $a_2$ |
|---|---|---|
| $b_1$ | 1 | 7 |
| $b_2$ | 7 | 1 |

(f)

| | $a_1$ | $a_2$ |
|---|---|---|
| $b_1$ | 1 | 7 |
| $b_2$ | 4 | 4 |

(g)

| | $a_1$ | $a_2$ |
|---|---|---|
| $b_1$ | 4 | 7 |
| $b_2$ | 4 | 1 |

(h)

| | $a_1$ | $a_2$ |
|---|---|---|
| $b_1$ | 7 | 1 |
| $b_2$ | 1 | 1 |

FIG. 10–4 Some potential outcomes of two factor experiments.

data, and in each case, decide whether or not there are likely to be main effects and/or an interaction. The answers are listed in Table 10-7.

TABLE 10-7 Effects present in Figure 10-4

| | *Main Effect of Factor A?* | *Main Effect of Factor B?* | *A × B Interaction* |
|---|---|---|---|
| (a) | No | No | No |
| (b) | Yes | No | No |
| (c) | No | Yes | No |
| (d) | Yes | Yes | No |
| (e) | No | No | Yes |
| (f) | Yes | No | Yes |
| (g) | No | Yes | Yes |
| (h) | Yes | Yes | Yes |

## TWO-FACTOR DESIGNS WITH MORE THAN TWO LEVELS OF ONE OR BOTH FACTORS

The strategy for analyzing and interpreting the results of a two-factor study is basically the same when more levels are added to one or both factors. Suppose we had included three Room-Size conditions (small, medium, and large) in the experiment on dormitory life. Several potential outcomes are given in Figure 10-5.

In Figures 10-5a and 10-5b, there are two main effects and an interaction. Ratings are affected by both room size and the number of roommates, and the magnitude of the difference in ratings of One- and Two-Roommates conditions depends upon room size. Figures 10-5c and 10-5d indicate similar main effects but no interaction (the lines are parallel). The difference between One- and Two-Roommates conditions is the same for all levels of the Room-Size variable.

Now we will add another level to the Number-of-Roommates variable (one, two, or three). In Figure 10-6, you will find several potential outcomes with three levels of both factors. Again, Figures 10-6a and 10-6b show interactions, whereas Figures 10-6c and 10-6d indicate the lack of an interaction.

## FACTORIAL NOTATION

Factorial designs of the sort we have been discussing are sometimes called $A \times B$ designs (read "A by B"). This means there are two factors or variables of interest (A and B). Any particular design can be specified by substituting the number of levels included for each variable. Thus, a $2 \times 2$ factorial includes two levels of Factor A ($a_1$ and $a_2$) and two levels of Factor B ($b_1$ and

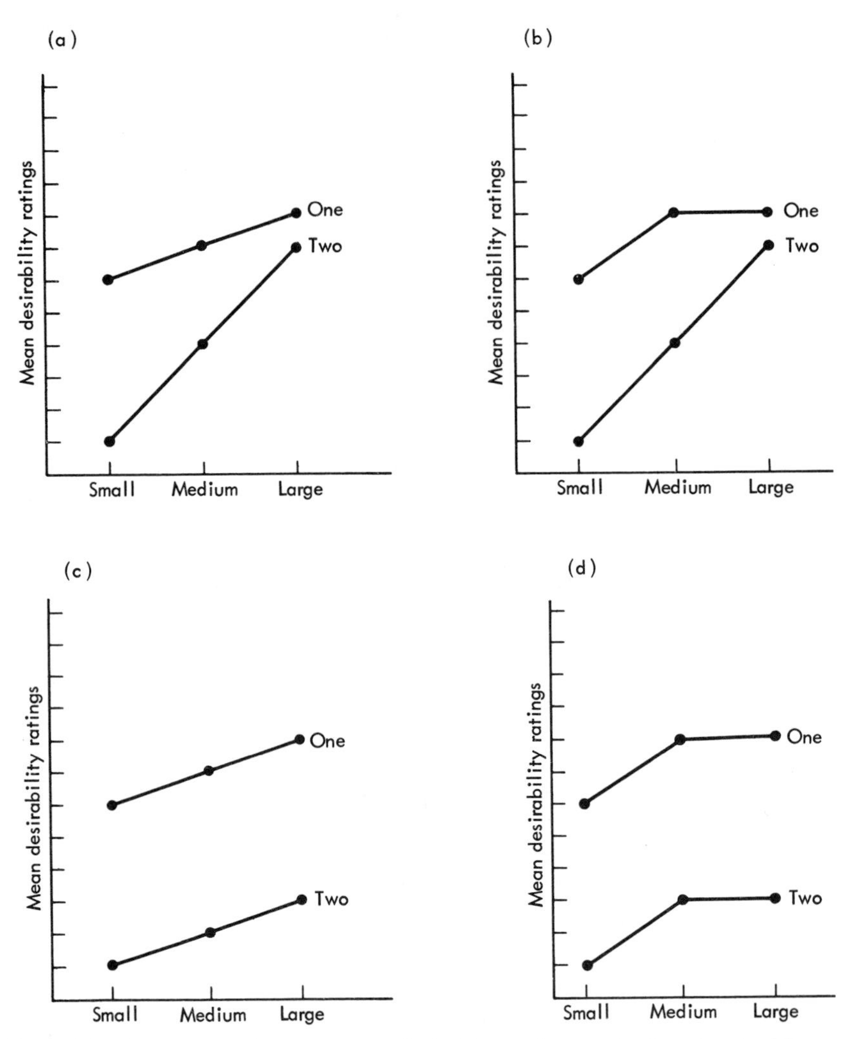


FIG. 10–5 Some potential outcomes of a factorial experiment with three levels of Factor A and two levels of Factor B.

$b_2$), and so forth. The total number of cells, or conditions, in a factorial experiment is always given by multiplying the number of levels of each factor ($2 \times 2 = 4$; $3 \times 2 = 6$; $3 \times 3 = 9$, and so forth).

The same notation extends to designs with more than two factors. For example, a $2 \times 3 \times 4$ factorial would be a design with three variables: two levels of Factor A, three levels of Factor B, and four levels of Factor C. The total number of conditions representing all combinations of treatments would be $2 \times 3 \times 4 = 24$. We will not discuss designs with more than two factors

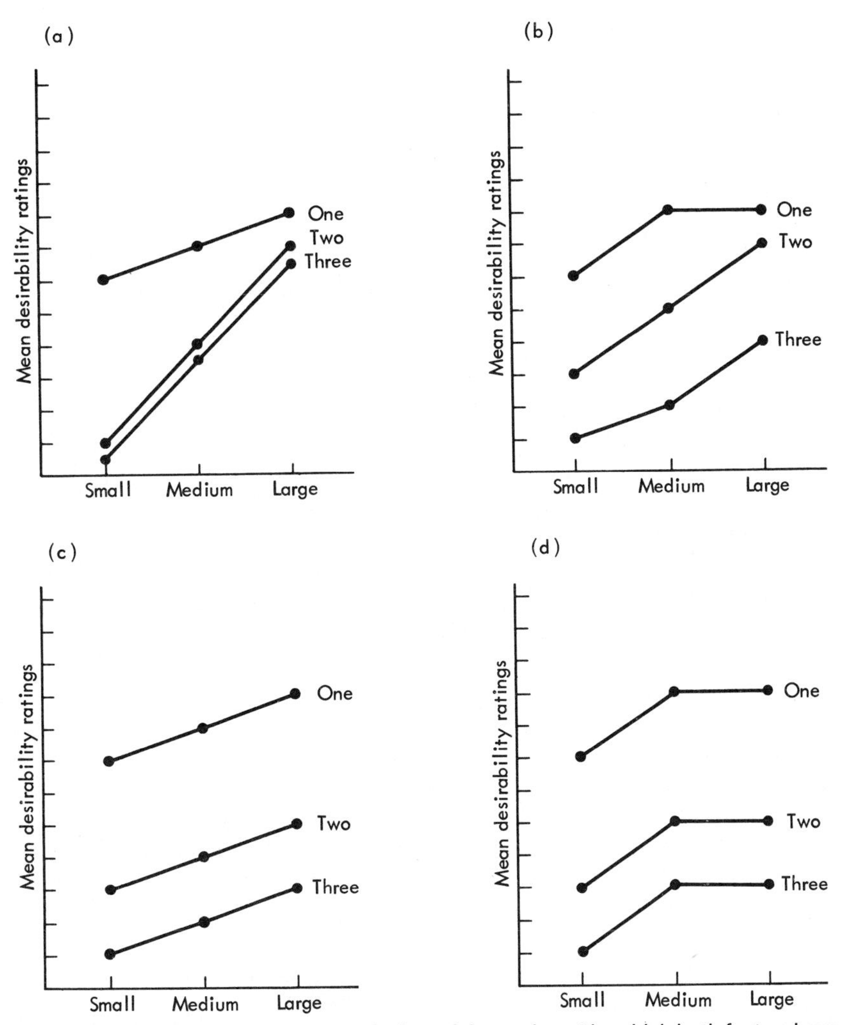


FIG. 10–6 Some potential outcomes of a factorial experiment in which both factors have three levels.

except to note that the principles involved in their statistical analysis and interpretation are extensions of those discussed in this chapter.

## COMPUTATIONAL PROCEDURES

In Chapter 8, we described how the total variability among the scores in a one-factor experiment could be thought of as consisting of two parts: (1) a sum of squares associated with between-groups variability and (2) a sum of squares associated with within-group variability. When the population means are all equal, these are both estimates of experimental error or random variability, and the $F$ ratio ($MS_{\text{Between}}/MS_{\text{Within}}$) should approximate a value of 1.00. When the population means are not equal (i.e., when there is an effect of the experimental factor), the magnitude of the between-groups sum of squares reflects the experimental effect in addition to random variability. Thus, the $F$ ratio will be larger than 1.00.

In a two-factor experiment, the between-groups variability consists of three parts: a sum of squares associated with the potential effects of Factor A ($SS_A$), a sum of squares associated with the potential effects of Factor B ($SS_B$), and a sum of squares associated with the potential effect of the combination of levels of Factor A with levels of Factor B ($SS_{AB}$). Each of these sums of squares will reflect random variability plus any experimental effects. The within-group sum of squares ($SS_{\text{Within}}$) is again simply a pooled estimate of random variability among subjects treated alike. Thus, the total sum of squares $= SS_A + SS_B + SS_{AB} + SS_{\text{Within}}$.

Refer to Table 10-8a. The $SS_A$ reflects differences among the means of the conditions of Factor A ($\bar{X}_{a_1}$, $\bar{X}_{a_2}$, $\bar{X}_{a_3}$, and so forth) summed across levels of Factor B. The $SS_B$ reflects differences among the means of the conditions of Factor B ($\bar{X}_{b_1}$, $\bar{X}_{b_2}$, $\bar{X}_{b_3}$, and so forth) summed across levels of Factor A. The $SS_{AB}$ reflects differences among the individual cell means ($\bar{X}_{a_1b_1}$, $\bar{X}_{a_1b_2}$, $\bar{X}_{a_1b_3}$, $\bar{X}_{a_2b_1}$, and so on) that are *not already accounted for* by the individual effects of Factor A or of Factor B alone.

### Computational Formulas for the Sums of Squares

As in the one-factor analysis of variance, the easiest computational formulas to work with are expressed in terms of group *totals* rather than group means. Refer to Table 10-8b for the following discussion.

First

$$SS_A = \frac{\sum (A)^2}{bn} - \frac{(\sum X_T)^2}{N}$$

The first term on the right equals the sum of the squared totals for each of the levels of Factor A divided by the number of scores summed prior to the squaring operation (the number of levels of Factor B combined for each level of Factor A $\times$ $n$) where $n$ is the number of observations in each cell. The second term equals the total of all the scores in the study squared and divided by the total number of scores.

Similarly

$$SS_B = \frac{\sum (B)^2}{an} - \frac{(\sum X_T)^2}{N}$$

TABLE 10-8 Detailed design of an experiment with three levels of each factor

(a) Cells Contain the Mean of Each Condition

| | *Factor A* | | | |
|---|---|---|---|---|
| *Factor B* | $a_1$ | $a_2$ | $a_3$ | |
| $b_1$ | $\bar{X}_{a_1b_1}$ | $\bar{X}_{a_2b_1}$ | $\bar{X}_{a_3b_1}$ | $\bar{X}_{b_1}$ |
| $b_2$ | $\bar{X}_{a_1b_2}$ | $\bar{X}_{a_2b_2}$ | $\bar{X}_{a_3b_2}$ | $\bar{X}_{b_2}$ |
| $b_3$ | $\bar{X}_{a_1b_3}$ | $\bar{X}_{a_2b_3}$ | $\bar{X}_{a_3b_3}$ | $\bar{X}_{b_3}$ |
| | $\bar{X}_{a_1}$ | $\bar{X}_{a_2}$ | $\bar{X}_{a_3}$ | $\bar{T}$ |

(b) Cells Contain the Total for Each Condition

| | *Factor A* | | | |
|---|---|---|---|---|
| *Factor B* | $a_1$ | $a_2$ | $a_3$ | |
| $b_1$ | $A_1B_1$ | $A_2B_1$ | $A_3B_1$ | $B_1$ |
| $b_2$ | $A_1B_2$ | $A_2B_2$ | $A_3B_2$ | $B_2$ |
| $b_3$ | $A_1B_3$ | $A_2B_3$ | $A_3B_3$ | $B_3$ |
| | $A_1$ | $A_2$ | $A_3$ | $\sum X_T$ |

$n$ = number of scores per cell
$N$ = total number of scores in the study

The first term on the right equals the sum of the squared totals for each of the levels of Factor B divided by the number of scores summed prior to the squaring operation (the number of levels of Factor A combined for each level of Factor B $\times$ $n$). The second term is exactly the same as the second term from $SS_A$.

Next

$$SS_{AB} = \frac{\sum (AB)^2}{n} - \frac{\sum (A)^2}{bn} - \frac{\sum (B)^2}{an} + \frac{(\sum X_T)^2}{N}$$

The first term on the right equals the sum of the squared totals for each condition or cell divided by the number of scores ($n$) summed prior to the squaring operation. The next two terms are the first term from $SS_A$ and the first term from $SS_B$, respectively. The last term is the same as the second term from both $SS_A$ and $SS_B$.

The pooled estimate of random variability is given by

$$SS_{\text{Within}} = \sum X^2 - \frac{\sum (AB)^2}{n}$$

The first term on the right is the sum of the squared values of each score, and the second term is the first term from $SS_{AB}$.

Finally

$$SS_{\text{Total}} = \sum X^2 - \frac{(\sum X_T)^2}{N}$$

This is the familiar computational formula for the sum of the squared deviations of each score from the grand mean $[\sum (X_i - \bar{T})^2]$.

These formulas are listed in the first column of Table 10-9. They are appropriate only when the number of observations or scores ($n$) is equal in all conditions of the study.

TABLE 10-9 Summary of computations for a two-factor experiment

| *Source* | *SS* | *df* | *MS* | *F* |
|---|---|---|---|---|
| Factor A | $\frac{\sum (A)^2}{bn} - \frac{(\sum X_T)^2}{N}$ | $a - 1$ | $\frac{SS_A}{df_A}$ | $\frac{MS_A}{MS_{\text{Within}}}$ |
| Factor B | $\frac{\sum (B)^2}{an} - \frac{(\sum X_T)^2}{N}$ | $b - 1$ | $\frac{SS_B}{df_B}$ | $\frac{MS_B}{MS_{\text{Within}}}$ |
| $A \times B$ | $\frac{\sum (AB)^2}{n} - \frac{\sum (A)^2}{bn} - \frac{\sum (B)^2}{an} + \frac{(\sum X_T)^2}{N}$ | $(a - 1)(b - 1)$ | $\frac{SS_{AB}}{df_{AB}}$ | $\frac{MS_{AB}}{MS_{\text{Within}}}$ |
| Within | $\sum X^2 - \frac{\sum (AB)^2}{n}$ | $(n - 1)(ab)$ | $\frac{SS_{\text{Within}}}{df_{\text{Within}}}$ | |
| Total | $\sum X^2 - \frac{(\sum X_T)^2}{N}$ | $N - 1$ | | |

In order to convert $SS_A$, $SS_B$, $SS_{AB}$, and $SS_{\text{Within}}$ to variance estimates (mean squares), each has to be divided by its *df*. These are listed in the second column of Table 10-9. The *df* associated with $SS_A$ is the number of levels of Factor A minus 1 ($a - 1$). Similarly, the *df* for $SS_B$ equals the number of levels of B minus 1 ($b - 1$). The *df* for $SS_{AB}$ equals the *df* for $SS_A$ times the *df* for $SS_B$ ($[a - 1][b - 1]$). The *df* for $SS_{\text{Within}}$ equals the number of scores in each condition minus 1, times the number of conditions $(n - 1)ab$. The total *df* equals the total number of observations in the study minus 1 ($N - 1$). And the *df* of $SS_A$, $SS_B$, $SS_{AB}$, and $SS_{\text{Within}}$ add up to the total *df*. The appropriate mean squares are indicated in the third column of Table 10-9.

In order to ask the three major questions that the two-factor design is intended to answer, the three $F$ ratios indicated in the last column of Table

10-9 are formed. Large values of $MS_A/MS_{\text{Within}}$ will lead us to conclude that there is a main effect of Factor A. Large values of $MS_B/MS_{\text{Within}}$ suggest a main effect of Factor B. Similarly, large values of $MS_{AB}/MS_{\text{Within}}$ indicate an interaction between Factors A and B. Obtained values of $F$ are compared with the appropriate sampling distribution of $F(df_{\text{numerator}}/df_{\text{denominator}})$ in Appendix E.

EXAMPLE 1

The data from Table 10-4 (the effect of room size and number of roommates on attitudes toward dormitory life) are reproduced in Table 10-10, along with some necessary preliminary calculations. The analysis appears in Table 10-11.

According to Appendix E, the critical value of $F(1, 12) = 4.75$ to reject the null hypothesis at the .05 level. Because each of our $F$ ratios exceeds this value, we may conclude there was significant effect of room size and of number of roommates and, in addition, an interaction of these two factors. From Figure 10-1, it is apparent that increasing the room size increased the desirability ratings more in the case of students living with two other people than in the case of students living with only one other person.

TABLE 10-10 Data from Table 10-4, showing preliminary calculation for the analysis of variance

| | *Small/One* | *Small/Two* | *Large/One* | *Large/Two* | $\sum$ | $(\sum X)^2$ |
|---|---|---|---|---|---|---|
| | 6 | 2 | 7 | 8 | | |
| | 5 | 1 | 8 | 7 | | |
| | 6 | 1 | 8 | 6 | | |
| | 7 | 0 | 9 | 7 | | |
| $\sum X$ | 24 | 4 | 32 | 28 | 88 | 7,744 |
| $\bar{X}$ | 6.00 | 1.00 | 8.00 | 7.00 | | |
| $\sum X^2$ | 146 | 6 | 258 | 198 | 608 | |
| $(AB)^2$ | 576 | 16 | 1,024 | 784 | 2,400 | |

Room Size

| *Number of Roommates* | *Small* | *Large* | *B* | $B^2$ |
|---|---|---|---|---|
| One | 24 | 32 | 56 | 3,136 |
| Two | 4 | 28 | 32 | 1,024 |
| *A* | 28 | 60 | | |
| $A^2$ | 784 | 3,600 | | |

TABLE 10-11 Analysis of variance for the data in Table 10-10

| *Source* | | *SS* | *df* | *MS* | *F* |
|---|---|---|---|---|---|
| Room size | $\frac{784 + 3{,}600}{8} - \frac{7{,}744}{16}$ | | | | |
| | $548 - 484$ | = 64 | 1 | 64 | 95.52 |
| Number of roommates | $\frac{3{,}136 + 1{,}024}{8} - 484$ | | | | |
| | $520 - 484$ | = 36 | 1 | 36 | 53.73 |
| Room size × Number of roommates | $\frac{576 + 16 + 1{,}024 + 784}{4} - 548 - 520 + 484$ | | | | |
| | $600 - 548 - 520 + 484$ | = 16 | 1 | 16 | 23.88 |
| Within | $608 - 600$ | = 8 | 12 | .67 | |
| Total | $608 - 484$ | = 124✓ | 15✓ | | |

TABLE 10-12 Data and preliminary calculations for a factorial experiment investigating the effects of drug dosage on activity level of hyperactive children

| | *Males* | | | | *Females* | | | | | |
|---|---|---|---|---|---|---|---|---|---|---|
| | *None* | *Low* | *Medium* | *High* | *None* | *Low* | *Medium* | *High* | $\sum$ | $(\sum X)^2$ |
| | 10 | 8 | 4 | 3 | 12 | 9 | 3 | 5 | | |
| | 11 | 7 | 3 | 4 | 8 | 6 | 6 | 2 | | |
| | 8 | 10 | 5 | 5 | 10 | 7 | 5 | 3 | | |
| | 7 | 9 | 7 | 2 | 9 | 5 | 2 | 1 | | |
| | 12 | 8 | 6 | 7 | 7 | 6 | 3 | 2 | | |
| | 4 | 5 | 5 | 1 | 5 | 4 | 4 | 4 | | |
| | 8 | 4 | 3 | 3 | 4 | 5 | 2 | 4 | | |
| | 6 | 7 | 2 | 1 | 5 | 6 | 3 | 2 | | |
| | 8 | 6 | 4 | 4 | 3 | 7 | 3 | 1 | | |
| | 9 | 8 | 4 | 2 | 8 | 8 | 5 | 1 | | |
| $\sum X$ | 83 | 72 | 43 | 32 | 71 | 63 | 36 | 25 | 425 | 180,625 |
| $\bar{X}$ | 8.30 | 7.20 | 4.30 | 3.20 | 7.10 | 6.30 | 3.60 | 2.50 | | |
| $\sum X^2$ | 739 | 548 | 205 | 134 | 577 | 417 | 146 | 81 | 2,847 | |
| $(AB)^2$ | 6,889 | 5,184 | 1,849 | 1,024 | 5,041 | 3,969 | 1,296 | 625 | 25,877 | |

| | *None* | *Low* | *Medium* | *High* | *B* | $B^2$ |
|---|---|---|---|---|---|---|
| Males | 83 | 72 | 43 | 32 | 230 | 52,900 |
| Females | 71 | 63 | 36 | 25 | 195 | 38,025 |
| *A* | 154 | 135 | 79 | 57 | 425 | |
| $A^2$ | 23,716 | 18,225 | 6,241 | 3,249 | | |

Analyze the data in Table 10-5, and verify that the $F$ values for the main effect of room size, number of roommates, and the room size by number of roommates interaction are 23.88, 149.25, and 0.00, respectively.

EXAMPLE 2

The data in Table 10-12 present the results of a hypothetical study of the effects of various dosages of a drug on activity level of hyperactive children. The design is a 4 × 2 factorial. Factor A is dosage (none, low, medium, and high), and Factor B is sex (male and female). Higher scores represent greater activity. The means are graphed in Figure 10-7, and the analysis is given in Table 10-13.

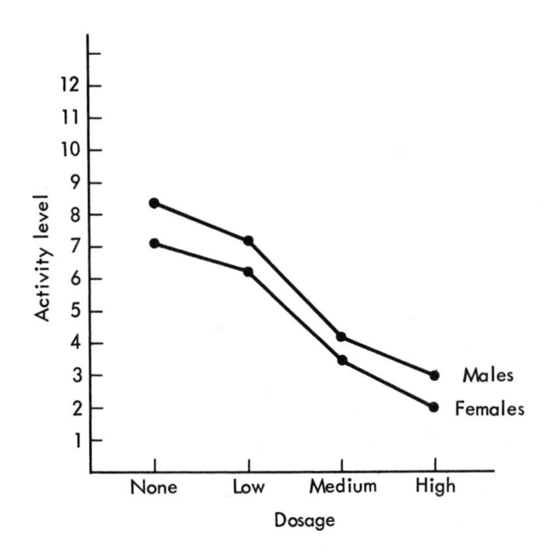


FIG. 10–7 Graphic representation of the means from Table 10-12.

TABLE 10-13 Analysis of variance for the data in Table 10-12

| *Source* | | *SS* | *df* | *MS* | *F* |
|---|---|---|---|---|---|
| Dosage | $\frac{23{,}716 + 18{,}225 + 6{,}241 + 3{,}249}{20} - \frac{180{,}625}{80}$ | | | | |
| | $2{,}571.55 - 2{,}257.81$ | = 313.74 | 3 | 104.58 | 29.05 |
| Sex | $\frac{52{,}900 + 38{,}025}{40} - \frac{180{,}625}{80}$ | | | | |
| | $2{,}273.12 - 2{,}257.81$ | = 15.31 | 1 | 15.31 | 4.25 |
| Dosage × sex | $\frac{25{,}877}{10} - 2{,}571.55 - 2{,}273.12 + 2{,}257.81$ | | | | |
| | $2{,}587.70 - 2{,}571.55 - 2{,}273.12 + 2{,}257.81$ | = 0.84 | 3 | 0.28 | $< 1$ |
| Within | $2{,}847 - 2{,}587.70$ | = 259.30 | 72 | 3.60 | |
| Total | $2{,}847 - 2{,}257.81$ | = 589.19✓ | 79✓ | | |

To evaluate the main effect of dosage, we need the critical value of $F(3, 72)$. The table in Appendix E does not include this particular $F$ distribution. However, Appendix E does give the critical value of $F(3, 70)$, which is 2.74 at the .05 level. The critical value of $F(3, 72)$ would actually be slightly less than 2.74.* Therefore, because our obtained $F$ was 29.05, we may conclude that dosage affected activity level. There was also a main effect of sex because the critical value of $F(1, 72)$ is less than the critical value of $F(1, 70)$, which is 3.98, and our obtained $F$ is 4.25. The interaction was not significant because our obtained $F$ was less than 1 and even with 3 and 1000 *df*, the critical value of $F$ is 2.61. In summary, increasing the dosage decreased activity level, females were less active than males, and increasing the dosage affected males and females similarly.

## Practice Problems

### A. A Thousand Words per Minute?

College sophomores were given a short course in speed-reading. They were randomly assigned to courses lasting 5, 15, or 25 sessions. Then, all the students were asked to read a passage and were given a test of comprehension. Within each session condition, a random one-third of the students were offered no money, one-third were offered $1.00, and one-third were offered $10.00, contingent on a certain level of performance. The investigator recorded the number of items correct on the comprehension test.

1. With 5 people per treatment condition, what would the total $N$ be?
2. Indicate by graph or table an outcome in which increasing the amount of money offered increased performance, in which increasing the number of sessions of the course increased performance, and in which the effect of increasing the amount of money offered did not depend upon how many sessions were given.
3. If you could add one additional *level* to this experiment, what would it be? Why?

### B. I Know What I Need

An experimenter used the following data to show that food preferences are in part determined by bodily requirements. Thirty rats were initially placed on a diet lacking mytoviacin. The rate of bar pressing was measured for each rat under one of the following three reward conditions: (1) food with no mytoviacin, (2) food with a moderate amount of mytoviacin added, and (3) food with a large amount of mytoviacin added. Because the rate of bar pressing increased with increasing amounts of mytoviacin, the experimenter concluded that deprivation can influence food preferences. Would you accept this conclusion? If not, why not? Specify a more appropriate design and a set of results that would support the investigator's hypothesis.

*Note that in Appendix E the critical value of $F$ for any given alpha level decreases as the *df* for the denominator (i.e., for $MS_{\text{within}}$) increases if everything else is equal. This is because $df_{MS_{\text{within}}}$ reflects the size of $n$; the larger $n$ is, the smaller the $F$ ratio required to obtain a significant effect.

## *C. Shaking the Weed*

Two smoking programs were compared for their effectiveness in inducing cigarette smokers to cut down or quit. Both programs involved 5 thirty-minute sessions, once a week for five weeks. At each session, the participants viewed a film. The films used in Program D emphasized the dangers and hazards of smoking (such as a close-up, in color, of the lungs of smokers); the films used in Program F emphasized the positive aspects of not smoking (such as nonsmokers having fun skiing and playing tennis). Because the investigator suspected that the relative effectiveness of the two programs might depend on the age of the smoker, age was included as a subject variable in the experiment. Group Y contained smokers aged 21 to 40, and Group 0 contained smokers aged 41 to 60. The people in Groups Y and 0 had smoked approximately the same number of packs per week prior to the study. A random half of each age-group was assigned to each program. The experimenter recorded the number of packs of cigarettes smoked by each person during the seventh week following the end of the program.

| Group Y(21–40) | | Group O(41–60) | |
|---|---|---|---|
| Group D | Group F | Group D | Group F |
| 3 | 1 | 0 | 10 |
| 7 | 0 | 0 | 5 |
| 5 | 4 | 1 | 5 |
| 7 | 4 | 6 | 7 |
| 5 | 0 | 2 | 5 |
| 5 | 1 | 3 | 5 |

Analyze these data. Were there significant main effects? Was there a significant interaction? Describe the outcome of the study in words.

## *D. Hubba and Hubba*

In a study of the relationship of water temperature and tank size to the sexual activity of the rare Hubba fish, 27 pairs of fish were randomly assigned to 27 fish tanks. There were three levels of water temperature (50°F, 60°F, and 70°F) and three levels of tank size (5, 10, and 15 gallons) combined factorially. The response measure was the number of times the male of each pair attempted to copulate with the female during an observation period. The scores are given below:

| 50°/5 gal | 60°/5 gal | 70°/5 gal | 50°/10 gal | 60°/10 gal | 70°/10 gal | 50°/15 gal | 60°/15 gal | 70°/15 gal |
|---|---|---|---|---|---|---|---|---|
| 1 | 2 | 1 | 3 | 3 | 4 | 2 | 6 | 5 |
| 0 | 0 | 1 | 1 | 2 | 3 | 3 | 4 | 4 |
| 1 | 0 | 1 | 2 | 3 | 2 | 1 | 2 | 3 |

Analyze these data, and describe the outcome of the study in words.

## *E. The Numbers Game*

In a study of the effects of eating and drinking on driving, 16 volunteers were randomly assigned to one of four conditions: 1 ounce of alcohol and a light snack; 1 ounce of alcohol and no snack; 2 ounces of alcohol and a snack; 2 ounces of alcohol and no snack. The number of errors on a driving test for each person was:

| 1 ounce/Snack | 1 ounce/No Snack | 2 ounces/Snack | 2 ounces/No Snack |
|---|---|---|---|
| 1 | 3 | 4 | 4 |
| 2 | 5 | 3 | 5 |
| 2 | 4 | 3 | 6 |
| 3 | 4 | 2 | 5 |

1. Analyze these data with a 2 × 2 analysis of variance.
2. Analyze the Amount-of-Alcohol variable with a one-factor analysis of variance (ignoring the Amount-of-Food variable).
3. Why is the main effect of amount of alcohol significant in the 2 × 2 analysis but not significant in the one-factor analysis of variance?

# Regression and Correlation

# 11

WILL A PARTICULAR APPLICANT SUCCEED academically in graduate, medical, or law school? Are two tall parents more likely than two average parents to have a tall child? Is a baby who is deprived of attention and affection early in life more likely to grow up with psychological problems than a baby who receives the usual amount of parental care? These three questions all ask for a *prediction*, which in each case could be made by knowing the relationship between the two characteristics, events, or experiences involved. We could answer the last question, for example, if we knew the exact relationship between parental care and psychological development.

This chapter is concerned with statistical procedures for describing and estimating the strength of relationships, usually for the purpose of deriving predictions from them. The specific techniques we will discuss provide the cornerstone for research-based predictions in many areas of government and industry as well as all basic and applied sciences. As the chapter title implies, these techniques are of two related types: *regression* and *correlation.*

## REGRESSION

There are circumstances in which we want to use the results of studies to make decisions about individual cases. Often, it is useful to predict a score for an individual given some prior information about him or her. A common case is the task of a personnel officer in an industrial setting. Should a certain individual be hired for a specific job? Some information about the candidate may be available, perhaps a score on a test of skills relevant to the job. In light of this information, the personnel officer must bet, or predict, how the person would work out. Vocational counselors give advice to students about the course of study in which they are likely to be successful, often using high school records and intelligence and interest test scores to decide what advice to give the student. Government officials may decide how much grain to sell to foreign countries after predicting the expected grain crop based on information about current weather conditions. The trick is, of course, to come up with the best prediction given the facts in hand. Even a blind guess is a prediction, but probably not a very good one.

### Logic of Prediction

The use of statistics as a tool in prediction involves evaluating the extent to which knowing an individual's or event's status on one variable ($X$) reduces uncertainty about its status on a second variable ($Y$). Suppose we asked you to predict or guess the high school grade-point average of Mary Smith. If you knew the mean and range of the distribution of grade-point averages in her class, what would be your best guess about Mary Smith's score? Your best guess would be the mean because in many distributions scores tend to cluster around the mean and therefore there is a higher probability that Mary Smith's score is near the mean than that it is near any other particular score in the distribution. You will almost certainly be in error about your prediction, of course, because Mary's score probably is not going to coincide with the mean, but choosing the mean would tend to minimize the degree of your error. If you chose the lowest score in the distribution and Mary's score happened to be the highest, you would have a very large error in prediction. A large potential error is avoided by choosing the mean, and in the last analysis, a correct prediction awaits anyone who can reduce to nil the size of all potential errors.

Now suppose we told you that we also had scores on reasoning problems for all the students, including Mary, as well as the grade-point averages of each. The question would then become: Can you improve your prediction of Mary's grade-point average by taking her score on the reasoning problems into account? To simplify the answer, let us assume there are only 6 people in the class. Their scores on the two variables (reasoning problems and grade-point average) are shown in Figure 11-1a. Figure 11-1b is a *scatterplot* of the same data. Usually, the variable that is being used to predict ($X$) is placed on the horizontal axis, and the variable to be predicted ($Y$) is placed on the vertical axis. Each point on the graph represents an individual's scores on the two variables.

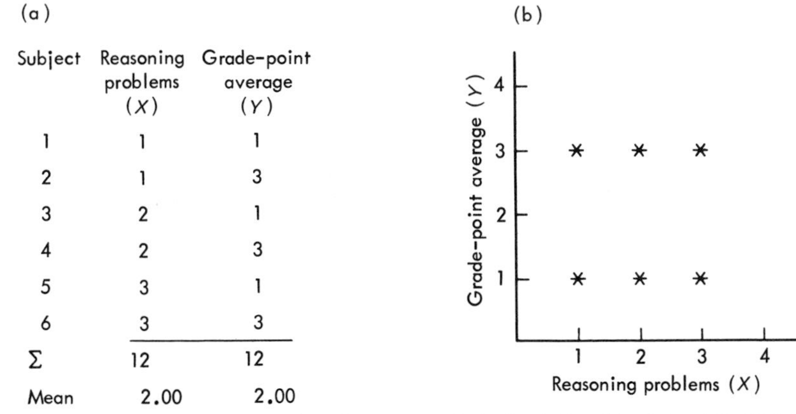

(a)

| Subject | Reasoning problems ($X$) | Grade-point average ($Y$) |
|---|---|---|
| 1 | 1 | 1 |
| 2 | 1 | 3 |
| 3 | 2 | 1 |
| 4 | 2 | 3 |
| 5 | 3 | 1 |
| 6 | 3 | 3 |
| Σ | 12 | 12 |
| Mean | 2.00 | 2.00 |



FIG. 11–1 Hypothetical data showing no statistical relationship between reasoning scores and grade-point averages for the 6 students in Mary Smith's class. The same data are presented in tabular form (a) and then as a scatterplot (b).

As it turned out, Mary's reasoning-problem score was 1.00. Does this information help you in your prediction of her grade-point average? No, because the distribution of possible grade-point scores for people who scored 1.00 on the reasoning test is as wide as it is for the whole class. When you knew only the total range of the $Y$ scores and not Mary's score on $X$, your best guess was the mean of $Y$, 2.00. Now, although you know her score on $X$, the same range of $Y$ scores still applies (1.00–3.00), and accordingly your best guess for Mary's grade-point average is still 2.00. Knowing Mary's reasoning score in this case did not reduce your uncertainty about her grade-point average.

Now consider the data in Figure 11-2. Here the grade-point averages for the class range from 0 to 4, and without any further information, your best guess about any particular score is still the mean of $Y$, or 2.00. However, if you know that Mary's score on the reasoning problems is 1.00, what have you gained? For people with an $X$ score of 1.00, the $Y$ scores range from 0 to 2.00. Your best guess for Mary's score is now the mean of the scores *of those people with an X score of* 1.00, in this case $(0 + 2)/2 = 1.00$. You have gained information by knowing Mary's $X$ score because the range of her possible $Y$ scores has been reduced from 0–4 to 0–2. You now have a better guess than simply the mean of the entire class, a guess that will reduce your potential error from a maximum error of 2 points to a maximum error of 1 point.

### Linear Rules for predicting $Y$ from $X$

In Figure 11-3, a line has been drawn through each of our previous scatterplots. In each case, the line is drawn through the points representing our best

(a)

| Subject | Reasoning problems ($X$) | Grade-point average ($Y$) |
|---|---|---|
| 1 | 1 | 0 |
| 2 | 1 | 2 |
| 3 | 2 | 1 |
| 4 | 2 | 3 |
| 5 | 3 | 2 |
| 6 | 3 | 4 |
| Σ | 12 | 12 |
| Mean | 2.00 | 2.00 |

(b)

Grade-point average ($Y$)

Reasoning problems ($X$)

FIG. 11–2 Hypothetical data showing a positive statistical relationship between reasoning scores and grade-point averages for the 6 students in Mary Smith's class. In contrast with the data presented in Figure 11–1, we would be better able to predict Mary's grade-point average in this case if we knew her score on the reasoning problems than if we did not.

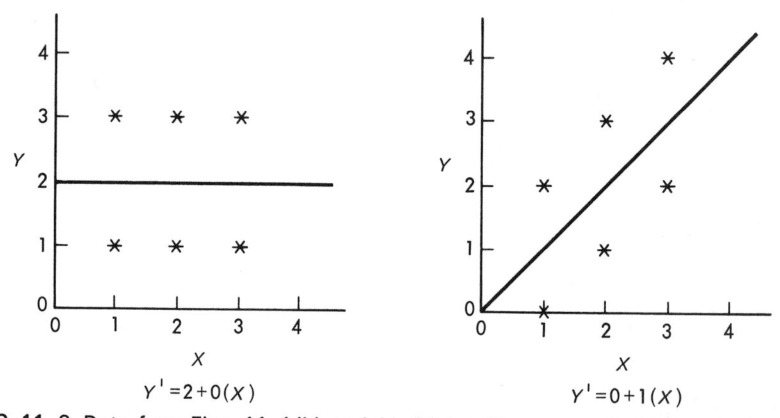


FIG. 11–3 Data from Figs. 11–1(b) and 11–2(b), with an equation in each case to predict $Y$ from $X$. Note that the predicted value of $Y$ is designated $Y'$.

guess about $Y$ for each value of $X$, $Y'$ (read "Y prime"). Each of these lines represents a rule for predicting values of $Y$ from information about $X$. As you may remember, a straight line can be described by an equation of the form

$$Y' = a + b(X)$$

where $a$ is the point at which the line crosses the vertical axis (the $Y$ intercept) and b is the slope of the line.

In the case of Figure 11-1, $a = 2.00$; and because the line is absolutely parallel to the horizontal axis, the slope, $b$, is 0. Thus, our predicted grade-point average is

$$Y' = 2 + 0(X)$$

Note also that values of $X$ will never change the value of $Y'$ when the slope is 0; we would therefore make the same prediction for $Y$ regardless of the person's score on $X$. A more technical way of describing this state of affairs is to say that there is no linear relationship between $X$ and $Y$.

This can be contrasted with our second example, in Figure 11-2. Here, as you should verify for yourself,

$$Y' = 0 + 1(X)$$

Because our predicted values of $Y$ change depending on the corresponding value of $X$, we would say technically that there is a linear relationship between the two variables. The contrast between the two sets of data is shown in Figure 11-3.

These were simple examples to illustrate the differences between situations in which $X$ does and does not give you information about $Y$ and to illustrate how a prediction rule might take the form of a linear equation. Ordinarily, we would have many more data points, and they would not fall into such an obvious and tidy pattern. For these cases, there is a mathematical way to determine the best-fitting prediction line (or *regression line*).

*Line of least squared error.* The line that we want is the line that will minimize our errors in prediction for the whole set of data. Errors in prediction are indicated on Figure 11-4. The distance between the actual values of $Y$ and our predicted values of $Y$ (that is, $Y'$) represents error in prediction. Thus, for each individual, the difference between the true value of $Y$ and our predicted value ($Y - Y'$) equals error. The prediction line that is chosen is the line for which the sum of the *squared* errors in prediction, $\sum (Y - Y')^2$, has the lowest value.

Although we will not present the mathematical proof here, it can be shown that the line $Y' = a + bX$, which minimizes the squared errors in prediction, has the following values for the $Y$ intercept ($a$) and slope ($b$):

$$a = \bar{Y} - b\bar{X} \quad \text{and} \quad b = \frac{\sum [(X - \bar{X})(Y - \bar{Y})]}{\sum (X - \bar{X})^2}$$

As you may have noticed, the denominator of the slope is the by-now-familiar sum of squares ($SS_X$, because the sum of squares is computed for the scores on the $X$ variable). The numerator is sometimes called the *sum of squares of the*

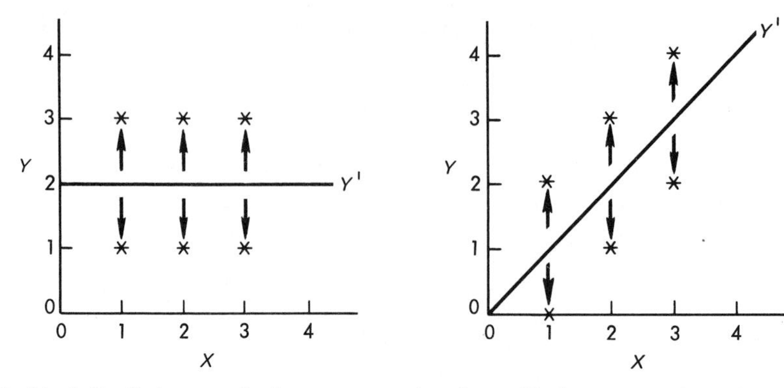


FIG. 11–4 Predictions made from a regression line will always contain a degree of error, even when there is a clear relationship between $X$ and $Y$. The little asterisks, designated by the arrows, can be thought of as some possible real values of $Y$, whereas our prediction in each case would have been the point on the regression line from which the arrow emanates. The data on the left are from Fig. 11–1(b), and those on the right are from Fig. 11–2(b).

*cross products*, or $SS_{XY}$. It is an index of the extent to which $X$ and $Y$ values for each subject vary together. Subjects whose deviation scores on both variables are large and *positive* will yield large positive values of

$$(X - \bar{X})(Y - \bar{Y}).$$

Similarly, people whose deviation scores on both variables are large and *negative* will also yield large positive values of $(X - \bar{X})(Y - \bar{Y})$. Hence, if the relative standing of the subjects tends to be the same on both variables, the quantity $\sum [(X - \bar{X})(Y - \bar{Y})]$ will have a large value. But if positive deviations on $X$ sometimes correspond to positive deviations on $Y$ and sometimes to negative deviations on $Y$, the resulting values of $(X - \bar{X})(Y - \bar{Y})$ will tend to cancel each other out, and the value of $\sum [(X - \bar{X})(Y - \bar{Y})]$ will tend toward 0.

The computational formula for $b$ is

$$b = \frac{SS_{XY}}{SS_X} = \frac{\sum XY - [(\sum X)(\sum Y)]/N}{\sum X^2 - [(\sum X)^2/N]}$$

EXAMPLE 1

We will now use these formulas to determine the regression line for the data in Figure 11-2. The values you need for the computations are given in Table 11-1.

TABLE 11-1 Data from Figure 11-2, including preliminary calculations necessary to determine the regression line

| Subject | *Reasoning Problems* $X$ | *Grade-point Average* $Y$ | $X^2$ | $Y^2$ | $XY$ |
|---|---|---|---|---|---|
| 1 | 1 | 0 | 1 | 0 | 0 |
| 2 | 1 | 2 | 1 | 4 | 2 |
| 3 | 2 | 1 | 4 | 1 | 2 |
| 4 | 2 | 3 | 4 | 9 | 6 |
| 5 | 3 | 2 | 9 | 4 | 6 |
| 6 | 3 | 4 | 9 | 16 | 12 |
| $\Sigma$ | 12 | 12 | 28 | 34 | 28 |
| mean | 2.00 | 2.00 | | | |

$$b = \frac{(\sum XY) - [(\sum X)(\sum Y)]/N}{\sum X^2 - [(\sum X)^2/N]}$$

$$= \frac{28 - [(12 \times 12)]/6}{28 - [(12 \times 12)]/6}$$

$$= \frac{28 - 24}{28 - 24}$$

$$= \frac{4}{4}$$

$$= 1$$

and

$$a = \bar{Y} - b\bar{X}$$

$$a = 2 - (1 \times 2)$$

$$a = 0$$

Therefore

$$Y' = 0 + 1(X)$$

For additional practice, verify that $Y' = 2 + 0(X)$ for the data in Figure 11-1.

EXAMPLE 2

The data in the first two columns of Table 11-2 represent the results of a hypothetical study of divorced people whose parents were also divorced. The two measures obtained from each person in a sample of 12 were the number of years they were married and the number of years their parents were

TABLE 11-2 Data from a study of divorced people whose parents were also divorced. Preliminary calculations shown are for determining the equation that best describes the regression line of the $X$ and $Y$ relationship

| | Length of Marriage (years) | | | | |
|---|---|---|---|---|---|
| | *Parents* | *Subject* | | | |
| *Subject* | $X$ | $Y$ | $X^2$ | $Y^2$ | $XY$ |
| 1 | 1 | 3 | 1 | 9 | 3 |
| 2 | 2 | 4 | 4 | 16 | 8 |
| 3 | 4 | 4 | 16 | 16 | 16 |
| 4 | 5 | 5 | 25 | 25 | 25 |
| 5 | 3 | 5 | 9 | 25 | 15 |
| 6 | 2 | 3 | 4 | 9 | 6 |
| 7 | 3 | 4 | 9 | 16 | 12 |
| 8 | 1 | 3 | 1 | 9 | 3 |
| 9 | 5 | 6 | 25 | 36 | 30 |
| 10 | 4 | 6 | 16 | 36 | 24 |
| 11 | 4 | 5 | 16 | 25 | 20 |
| 12 | 2 | 5 | 4 | 25 | 10 |
| $\Sigma$ | 36 | 53 | 130 | 247 | 172 |
| mean | 3.00 | 4.42 | | | |

married. What regression line best describes the relationship between these two measures?

The preliminary calculations you need to compute the slope and $Y$ intercept of the regression line are also given in Table 11-2. They have been substituted into the appropriate formulas below.

$$b = \frac{(\sum XY) - [(\sum X)(\sum Y)]/N}{\sum X^2 - [(\sum X)^2/N]}$$

$$= \frac{172 - (36 \times 53)/12}{130 - (36^2/12)}$$

$$= \frac{172 - 159}{130 - 108}$$

$$= \frac{13}{22}$$

$$= 0.59$$

$$a = \bar{Y} - b\bar{X}$$

$$= 4.42 - (0.59 \times 3.00)$$

$$= 4.42 - 1.77$$

$$= 2.65$$

Therefore

$$Y' = 2.65 + 0.59X$$

This is the prediction line that minimizes the squared prediction errors for this set of data.

Now, using this information, what would be your best guess about the length of a person's marriage if you know his or her parents had been married 2 years?

$$\begin{aligned} Y' &= 2.65 + 0.59X \\ &= 2.65 + (0.59 \times 2) \\ &= 2.65 + 1.18 \\ &= 3.83 \end{aligned}$$

A scatterplot of these data, with the appropriate regression line indicated, is given in Figure 11-5. In order to construct the regression line, solve the regression equation for any two points, and connect these with a straight line.

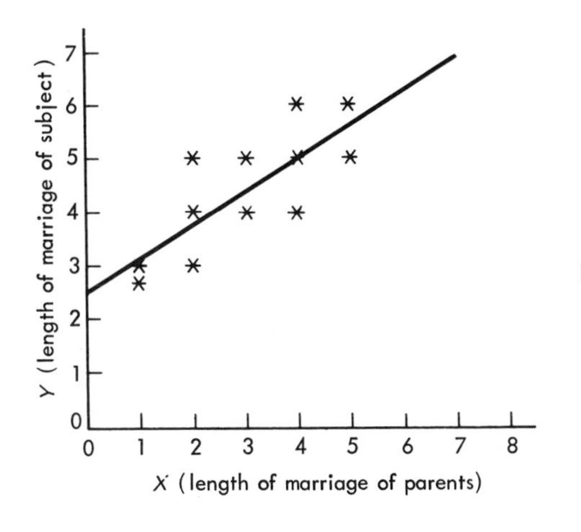


FIG. 11–5 Data from Table 11-2, shown as a scatterplot. Note how the actual values vary around the regression line considerably, telling us the degree of relationship is far from perfect.

## CORRELATION

There is an important question to which we must return now, and it has to do with the *margin of error* in these regression-line predictions. We know that a regression equation will give us a specific prediction of $Y$ from $X$. But how accurate is that prediction? Although in Example 2 we predicted that a person whose parents were married 2 years would himself remain married for 3.83 years, we certainly did not think that that prediction would be exactly right—not very often, at least. The accuracy of our prediction will depend on how closely $X$ and $Y$ vary together or are related, and this is the essential question behind the statistical concept of *correlation*.

To explore the meaning of correlation further, refer to the two sets of data in Table 11-3. Verify that the best-fitting regression line in both cases is $Y' = 1.2 + 0.70X$. The data are plotted in Figure 11-6. Notice that although both sets of data have the same regression equation, the data cluster more tightly around the regression line in Set A than they do in Set B. One consequence of this tighter fit of the line to the data points is that we would expect smaller errors in prediction in the case of Set A compared with Set B. The relationship between $X$ and $Y$ seems stronger; the two variables are more closely correlated in the case of Set A.

It would be useful to have an index of the strength of the relationship between two variables that reflects the difference between these two sets of

TABLE 11-3 Two sets of data with the same regression line and different correlation coefficients

| | | | Set A | | |
|---|---|---|---|---|---|
| | $X$ | $Y$ | $X^2$ | $Y^2$ | $XY$ |
| | 0 | 1 | 0 | 1 | 0 |
| | 0 | 1 | 0 | 1 | 0 |
| | 2 | 3 | 4 | 9 | 6 |
| | 2 | 3 | 4 | 9 | 6 |
| | 4 | 4 | 16 | 16 | 16 |
| | 4 | 4 | 16 | 16 | 16 |
| | 6 | 5 | 36 | 25 | 30 |
| | 6 | 5 | 36 | 25 | 30 |
| | 8 | 7 | 64 | 49 | 56 |
| | 8 | 7 | 64 | 49 | 56 |
| $\Sigma$ | 40 | 40 | 240 | 200 | 216 |
| $\bar{X}$ | 4.0 | 4.0 | | | |

| | | | Set B | | |
|---|---|---|---|---|---|
| | $X$ | $Y$ | $X^2$ | $Y^2$ | $XY$ |
| | 0 | 0 | 0 | 0 | 0 |
| | 0 | 2 | 0 | 4 | 0 |
| | 2 | 2 | 4 | 4 | 4 |
| | 2 | 4 | 4 | 16 | 8 |
| | 4 | 3 | 16 | 9 | 12 |
| | 4 | 5 | 16 | 25 | 20 |
| | 6 | 4 | 36 | 16 | 24 |
| | 6 | 6 | 36 | 36 | 36 |
| | 8 | 6 | 64 | 36 | 48 |
| | 8 | 8 | 64 | 64 | 64 |
| $\Sigma$ | 40 | 40 | 240 | 210 | 216 |
| $\bar{X}$ | 4.00 | 4.00 | | | |

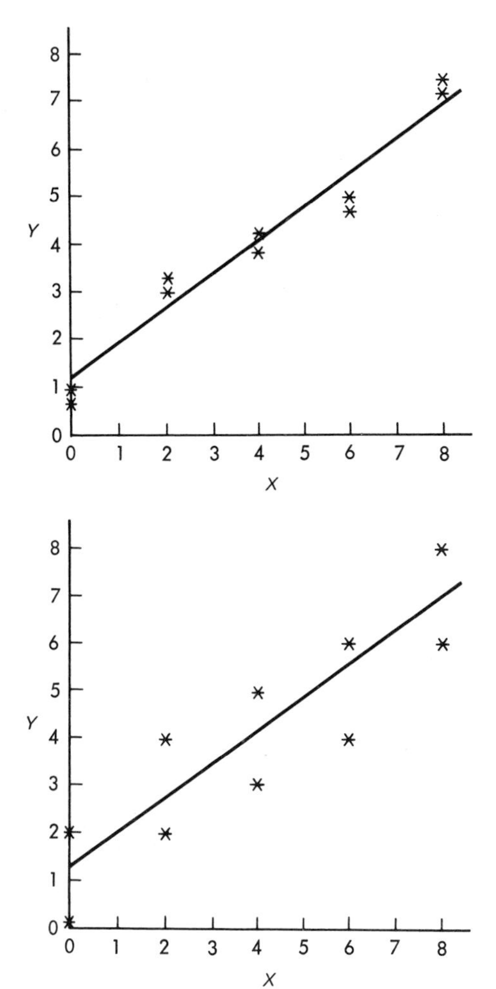


FIG. 11–6 Data from Table 11-3 (same regression line and different correlation coefficients).

data. The *Pearson correlation coefficient* ($r$) is just such an index. The computational formula for $r$ is as follows:

$$r = \frac{SS_{XY}}{\sqrt{SS_X SS_Y}}$$

$$= \frac{(\sum XY) - [(\sum X)(\sum Y)]/N}{\sqrt{[\sum X^2 - (\sum X)^2/N][\sum Y^2 - (\sum Y)^2/N]}}$$

By substituting the relevant preliminary calculations from Table 11-3 into this formula, the correlation coefficient ($r$) for each set of data may be obtained:

For Set A

$$r_A = \frac{216 - (40 \times 40)/10}{\sqrt{[240 - (40^2/10)][200 - (40^2/10)]}}$$

$$= \frac{216 - 160}{\sqrt{(240 - 160)(200 - 160)}}$$

$$= \frac{56}{\sqrt{3{,}200}}$$

$$= \frac{56}{56.57}$$

$$= .99$$

For Set B

$$r_B = \frac{216 - (40 \times 40)/10}{\sqrt{[240 - (40^2/10)][210 - 40^2/10)]}}$$

$$= \frac{216 - 160}{\sqrt{(240 - 160)(210 - 160)}}$$

$$= \frac{56}{\sqrt{4{,}000}}$$

$$= \frac{56}{63.24}$$

$$= .88$$

As you can see, although both sets of data have the same best-fitting regression line, the correlation between $X$ and $Y$ for Set A (.99) is larger than the correlation between $X$ and $Y$ for Set B (.88), reflecting the stronger relationship between $X$ and $Y$ in the data for Set A.

In general, $r$ provides information about both the direction and the magnitude of a relationship. Direction is indicated by the sign of $r$ (+ or −). When a positive $r$ is obtained, $X$ and $Y$ are positively related; high scores on $X$ tend to be associated with high scores on $Y$, and low scores on $X$ tend to go along with low scores on $Y$. A child's age and his grade in school are positively related; the older he is, the higher his grade in school is likely to be. A negative $r$ tells us that high scores on $X$ tend to be associated with low scores on $Y$ and that low scores on $X$ tend to be associated with high scores on $Y$. Experience and mistakes are negatively correlated; the more experience people have at a task, the fewer mistakes they tend to make.

The correlation coefficient, $r$, may range from +1.00, which indicates a perfect positive relationship, to −1.00, which indicates a perfect negative relationship. When $r$ is 1.00 (+ or −), either variable can be predicted perfectly from the other. As $r$ decreases in absolute value from 1.00, moving closer to 0.00, the ability to predict one variable from the other decreases.

When $r$ is 0.00, $X$ and $Y$ are unrelated and knowledge of one would provide no help in predicting the other. The $r$ between $X$ and $Y$ in Figure 11-1 (page 161) is 0.00, and that is the reason we were unable to predict Mary Smith's score in that example. Furthermore, in relation to our earlier discussion of regression, when $r = 0.00$, the slope of the regression line will also be 0. These relationships may all be seen in Figure 11-7; note particularly that the

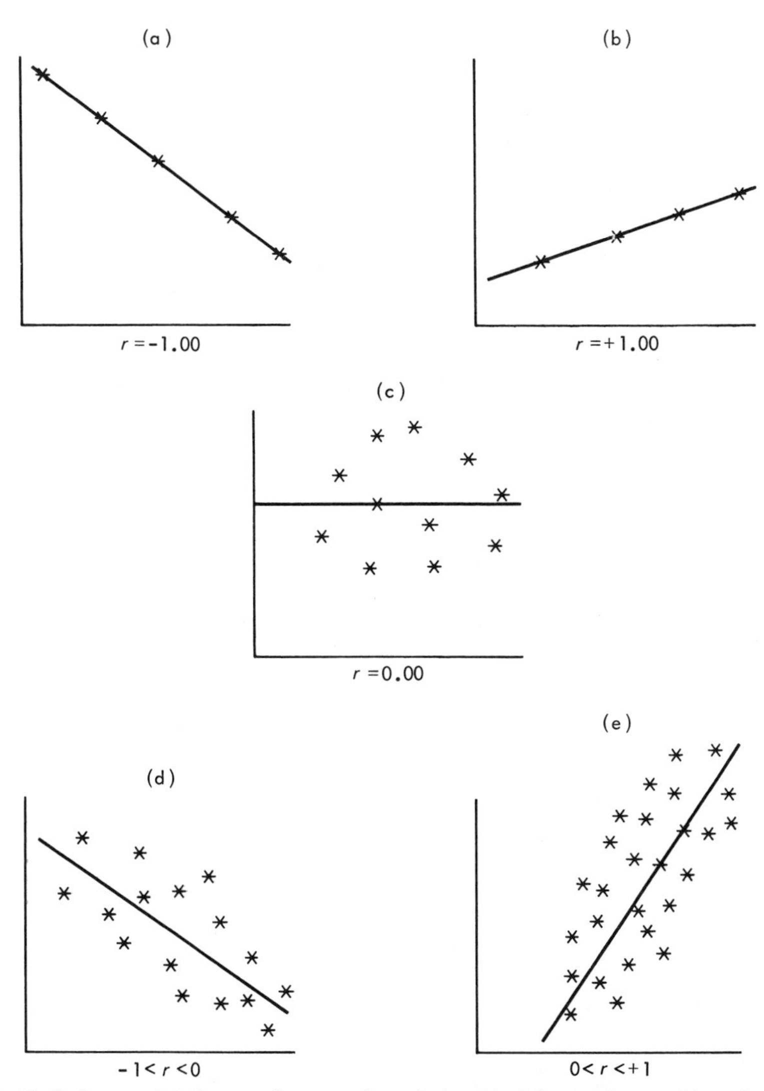


FIG. 11–7 Scatterplots for a perfect negative relationship (a), a perfect positive relationship (b), a nil relationship (c), and moderate negative and positive relationships (d and e).

larger our absolute value of $r$, the less variation there is about the regression line and therefore the smaller our errors in prediction will be.

### Testing the Hypothesis that $\rho = 0$

In predicting Mary Smith's grade-point average, the only population in which we were interested was Mary's class, and we had direct information on all the pupils in it. Often, however, an investigator will be interested in a population far larger than anyone would care to study completely and thus will observe only a random sample of cases (obtaining measures of $X$ and $Y$ for each); the investigator will then draw an inference from the sample to the population, an idea with which you are already familiar. We shall examine the logic of this important use of $r$ in some detail.

If it were possible to measure every individual in the population and compute the correlation between $X$ and $Y$ measures, we would have the exact value of the population correlation coefficient, $\rho$ (the Greek letter *rho*). Thus, $r$ is an estimate of $\rho$, just as $\bar{X}$ is an estimate of $\mu$. Even if $\rho = 0$, we would sometimes expect, by chance, to observe values of $r$ that are greater or less than 0. When $\rho$ is in fact equal to 0, the sampling distribution of $r$ (for successive random samples of size $N$) has a mean of 0 and a standard deviation of $\sqrt{(1 - r^2)/(N - 2)}$. Thus, the obtained $r$ can be expressed as a standard score by forming the following ratio:

$$\frac{r - \rho}{\sqrt{(1 - r^2)/(N - 2)}}$$

Because this ratio is distributed as $t$ with $N - 2$ *df*, the hypothesis $H_0: \rho = 0$ can be tested by comparing the standard score with the appropriate $t$ distribution.

For example, consider the data in Table 11-2. Is there a correlation between the number of years parents were married and the number of years their children were married? In order to test $H_0: \rho = 0$, we first must obtain the value of $r$ for our sample of 12 cases:

$$\begin{aligned} r &= \frac{SS_{XY}}{\sqrt{SS_X SS_Y}} \\ &= \frac{\sum XY - [(\sum X)(\sum Y]/N}{\sqrt{[\sum X^2 - (\sum X)^2/N][\sum Y^2 - (\sum Y)^2/N]}} \\ &= \frac{172 - (36 \times 53)/12}{\sqrt{[130 - (36^2/12)][247 - (53^2/12)]}} \\ &= \frac{13}{\sqrt{22 \times 12.92}} \\ &= +.77 \end{aligned}$$

Is .77 sufficiently larger than 0 to reject the hypothesis $\rho = 0$? If we limit our probability of a Type I error (rejecting $H_0$ when it is in fact true) to .05, then our obtained value of $r$, expressed as a $t$ score, must exceed 2.228 (the critical value for $t$ with $df = N - 2$). To convert our obtained $r$ to a standard score,

$$\begin{aligned} t &= \frac{r - 0}{\sqrt{(1 - r^2)/(N - 2)}} \\ &= \frac{.77}{\sqrt{(1 - 0.77^2)/(12 - 2)}} \\ &= \frac{.77}{\sqrt{(1 - .59)/10}} \\ &= \frac{.77}{\sqrt{0.04}} \\ &= \frac{.77}{0.20} \\ &= 3.85 \end{aligned}$$

Because 3.85 is greater than 2.228, we may reject $H_0$ and conclude that there is a relationship between the length of divorced individuals' marriages and the length of the marriages of their parents.

## INTERPRETING *r*

Like any other estimate of a population value, the meaning of $r$ depends upon the methods by which it has been obtained and the thought that goes into its interpretation. There are several specific points that you should always keep in mind.

### Misleading *r*'s

The Pearson correlation coefficient and the regression line reflect and measure linear relationships between variables, but they are inappropriate for identifying *nonlinear* relationships and are often insensitive to curvilinear distributions in particular. Figure 11-8, for example, shows a possible relationship between anxiety and performance and certainly tells us that one of these variables is predictable in part from the other; people who are very high *or* very low in anxiety will perform more poorly than people who are moderately anxious. It would not be appropriate to test or describe such a relationship with $r$, and if we were to do so, the obtained value would be very near 0.00. Other correlational techniques (which will not be discussed here) are appropriate for these situations.

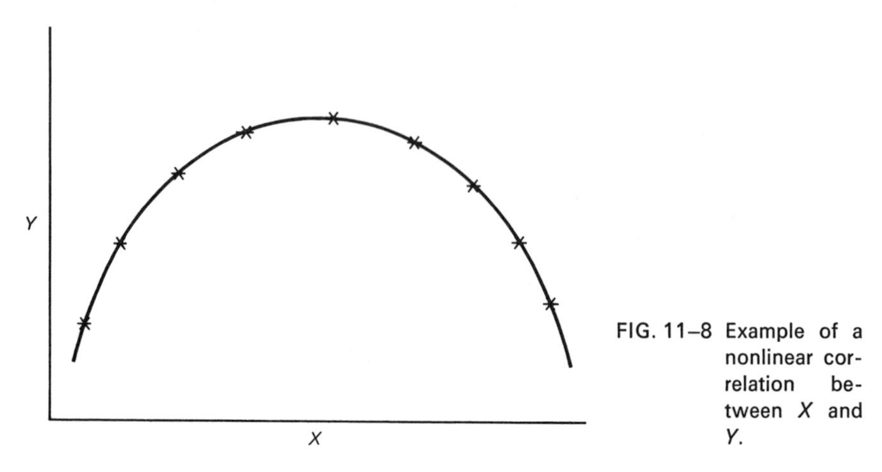


FIG. 11–8 Example of a nonlinear correlation between $X$ and $Y$.

In addition, even where the true relationship between two variables is linear, the value of $r$ will depend upon the range of cases observed (see Figure 11-9). If your sample is taken only from the middle range of scores for both variables, your obtained value of $r$ will generally be lower than the true $\rho$. Similarly, if your sample includes primarily cases from the extremes of the population, your obtained $r$ will be inflated. Thus, random sampling from the population (which, as you will recall, should yield a sample containing values in proportion to their occurrence in the population) is critical.

### Comparing $r$'s

Because correlation coefficients range in absolute value from 0.00 to 1.00, people are sometimes tempted to view them as percentages (i.e., when multiplied by 100) and then to assume, for example, that an $r$ of .50 reflects a relationship twice as strong as an $r$ of .25. But $r$'s are *not* percentages, and any such comparisons would be wrong. It can be proved (although we will not prove it here) that the appropriate rule of thumb in comparing correlations for relative strength is to compare *squared* $r$'s. An $r$ of .60 therefore has *four times* the predictive power of an $r$ of .30 because $(.60)^2 = .36$ and $(.30)^2 = .09$; $.36/.09 = 4.00$. When two $r$'s have been calculated from sample data and the investigator wishes to determine whether the population $\rho$'s are actually different, a statistical test of significance will be required. Such techniques are available, and although we will not detail them here, it should be apparent that they will be based on a logic similar to the one used for testing the difference between the two means.

### Correlation and Causality

Suppose an investigator drew a sample of 100 students and randomly assigned one of them to each of 100 different incentive conditions for chop-

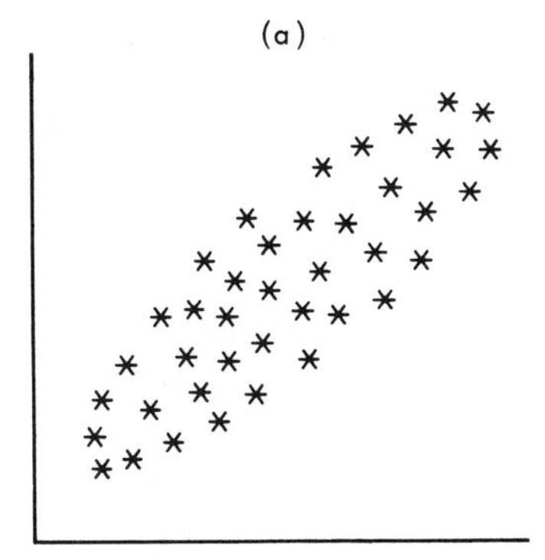


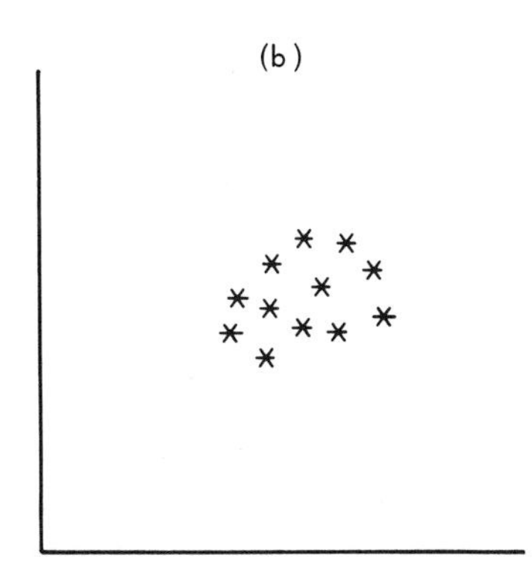


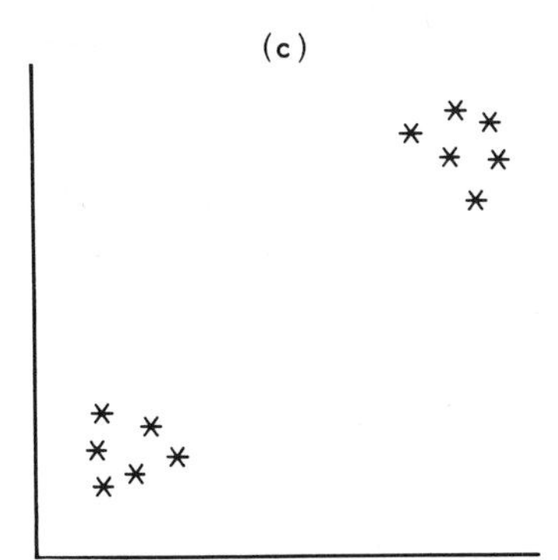


FIG. 11–9 Consequences for $r$ of sampling from either the middle or the extremes of the distribution.

ping wood, in even \$.50 increments from \$.50 to \$50.00 per log. We would then have an $X$ score and a $Y$ score for each student: the amount we promised to pay the student per log and the number of logs the student subsequently chopped, respectively. We might compute an $r$ for these data, and if it were significantly greater than 0.00, we could conclude that increasing incentives *caused* an increase in wood chopping. We would be able to say *caused* by virtue of having manipulated or directly controlled the $X$ variable, amount of incentive offered. However, $r$ is rarely used in this way. Instead, correlations are most often computed to determine the relationship that is already present between two variables, when neither has been manipulated. When data are collected and analyzed in this way, cause-and-effect relationships usually *cannot* be identified with confidence. As we have mentioned before, this fact is of extreme importance because ignorance of it in research can lead to absurd conclusions, some less easy to see through after the fact than the ones we will offer.

There is, for example, a high positive correlation between the number of churches in a city and the number of crimes committed in that city; the more churches a city has, the more crimes are committed. But this does not mean either that religion causes crime or that engaging in criminal behavior makes people more religious. Rather a third variable, population size, accounts for the relationship. The more people a city has living in it, the more churches and the more criminals it will have. Or consider the high positive correlation between the number of drownings on any given day and the consumption of ice cream on that day. Here, too, a third variable, temperature, is responsible for the relationship. As the weather gets warmer, more people go swimming (and thus more people drown) and more ice cream is eaten. Even when we are sure that two variables are causally related, we may not be able to be sure about the direction of the relationship. A strong positive correlation between the amount of praise given by parents and the school performance of their children may mean that parental praise stimulates good school performance or that good school performance stimulates parental praise.

The fact that we cannot infer the nature and direction of a causal relationship in these situations certainly does not mean that a cause-and-effect relationship does not exist; it merely means that in the absence of direct manipulation (i.e., doing a true experiment) correlations usually cannot give us solid information about causal events.

## Practice Problems

### *A. Like, Wow*

A psychologist suspects that LSD affects the speech center in the brain. Specifically, he believes that repeated use of LSD reduces a person's ability to retrieve verbal information. To see if he can obtain any evidence regarding this hunch, the psychologist advertises for paid subjects who have taken LSD at least once. Nine people of comparable IQ and education are selected from the applicants. All subjects are given a 50-item test.

Each item consists of a definition of a low-frequency English word; the subject's task is to produce the target word. Sample items might be:

| Definition | Target Word |
|---|---|
| to make things thinner or weaker by the addition of water | dilute |
| patronage bestowed in consideration of family relationship and not merit | nepotism |

Now here are the relevant data.

| Subject | Number of reported LSD trips | Number of *failures* to produce target word (errors) |
|---|---|---|
| 1 | 6 | 6 |
| 2 | 7 | 11 |
| 3 | 4 | 6 |
| 4 | 2 | 0 |
| 5 | 10 | 13 |
| 6 | 1 | 2 |
| 7 | 3 | 1 |
| 8 | 8 | 9 |
| 9 | 6 | 9 |

1. Find $r$.
2. Test the hypothesis that the population correlation, $\rho$, is 0.
3. Find the regression equation for predicting the test score (number of errors) from the number of LSD trips.
4. What would you conclude from this study?

## B. Snow on the Screen

A T. V. network's director of social research was interested in showing that children's programs with "heart" (i.e., featuring warm and kindly interactions between people) will teach children to be warm and kindly themselves. The investigator interviewed a sample of several thousand boys and girls from all over the United States that closely matched the U.S. population in age and geographic and economic background. The purpose of the interview was to find out how many hours of television each watched during the average week. Teachers also recorded the number of cooperative, helpful, or friendly acts performed by each child for a week. An $r$ was then computed to determine the correlation between hours of viewing and frequency of cooperative, helpful, or friendly acts. Do you think the $r$ will be significant? Why? How would you interpret an $r$ that did not differ significantly from 0?

## C. The Plot Thickens

The following scores for variables $X$ and $Y$ were obtained from two different groups of subjects. Produce a separate scatterplot for each, and indicate whether it suggests no correlation, a moderate correlation, or a high correlation. In each case, decide whether you think the $r$ will be positive or negative, and point out anything else you can see in the plot.

1.

| Subject | $X$ | $Y$ |
|---|---|---|
| 1 | 3 | 2 |
| 2 | 4 | 9 |
| 3 | 5 | 8 |
| 4 | 6 | 1 |
| 5 | 7 | 3 |
| 6 | 5 | 5 |
| 7 | 8 | 7 |

2.

| Subject | $X$ | $Y$ |
|---|---|---|
| 1 | 4 | 5 |
| 2 | 3 | 5 |
| 3 | 2 | 3 |
| 4 | 6 | 5 |
| 5 | 7 | 9 |
| 6 | 3 | 3 |
| 7 | 8 | 8 |

### *D. The Numbers Game*

In Problem A, increase the number of errors every subject made by 5. Now compute $r$ again, using the new scores. How does your new value compare with your old value? What do you think happened?

# Nonparametric Statistics: Dealing with Frequencies and Ranks

# 12

MOST READERS WILL HAVE GUESSED that the preceding chapters did not exhaust all the statistical procedures and designs used by social scientists. In fact, there are literally hundreds of statistical tests that could be described, each particularly useful in some special situation. Yet, in another sense, we have covered most of the basic ground on which statistical analysis rests; the more advanced techniques involve complex computations to be sure, but they rely on an underlying logic that is by now familiar. In this final chapter, we are therefore not concerned with presenting the sophisticated extensions of statistical reasoning that are available in advanced texts. Rather, we will describe several other basic statistical procedures that deviate somewhat from the pattern of those we have dealt with so far. We are speaking of the so-called nonparametric statistical procedures.

## NONPARAMETRIC STATISTICS

For the most part, modern statistical tests rely on a set of mathematical assumptions about the populations from which sample scores were drawn. The mathematical proofs underlying many statistical tests assume, for example, that the scores in the populations are distributed normally and have equal variances. These assumptions are in fact required by the logic of $t$ and $F$ tests. Occasionally, though, these assumptions are either highly dubious or flatly inapplicable to the data that are available. In such instances, investigators may select an appropriate *distribution-free*, or *nonparametric*, test.

The term nonparametric derives from the fact that population values are technically referred to as *parameters*. Therefore, statistical tests that require us to make certain assumptions (e.g., $\sigma_1^2 = \sigma_2^2$) about the population parameters from which our samples have been drawn are referred to as *parametric* statistical tests. Because these tests have great power (i.e., they minimize the probability of a Type II error), investigators will often use them even if certain assumptions have been violated to a degree. This tactic is not as risky as it may seem at first because statisticians have demonstrated that correct inferences often can be drawn even in the face of moderately severe violations.

In other situations, however, we cannot be certain about the effects that violating mathematical assumptions will have on the accuracy of our inferences. To provide tools for dealing with these situations, statisticians have devised an entire set of statistical procedures that do not require as many assumptions about the underlying distributions. These procedures are distribution-free in the sense that they are free of certain assumptions regarding parameters of the underlying populations. They have been given the general name *nonparametric statistics*.

Generally, it can be said that some nonparametric test is available as a substitute for a parametric test in any situation. However, although it is safer in a theoretical sense to use nonparametric tests whenever one is in doubt about one's assumptions, nonparametric tests are usually less powerful than their parametric counterparts, and so many investigators will resort to them only when absolutely necessary. Nonparametric tests *are* absolutely necessary when we deal with two types of data: categorical frequencies and ranks. We will therefore describe the nature and application of several of the more common nonparametric tests in these situations.

## DEALING WITH CATEGORICAL FREQUENCIES

If you ask a person whether he voted in the last presidential election, he will merely say "yes" or "no" or, perhaps, "I don't remember." Similarly, the investigator who wants to determine whether middle-aged people lie about their ages will be able to say of his or her research participants either that they did or did not lie, but there will be no true scores in the sense that one has an IQ or some other test score which might vary widely from one individual to another. Even when data do vary along a continuum (e.g., as they do in an examination of the number of years of education people have received), it is sometimes useful to divide the distribution into categories instead (e.g., college graduate, some college, and no college). In all these instances, we assign a small number of categories rather than numbers to the events of interest and ask about the frequency with which events occur rather than about their size or magnitude. Special statistical techniques are required for dealing with frequencies, and we will provide an introduction to them in this section.

### Concept of Expected Frequencies

Whenever a sample of data is collected in such a way that each observation is assigned to a category, *frequency data* are involved. The problem then becomes one of comparison and inference, and the solution also parallels a logic we have discussed earlier. Investigators working with frequencies must have a theoretical distribution against which the data in their obtained sample can be compared. Here, though, the theoretical distribution becomes the expected frequency with which observations would fall into each category used under a given set of circumstances. The sample data therefore provide an obtained set of frequencies, and statistical inferences will involve testing to determine whether the expected and obtained frequencies differ by a wider margin than would be expected at some predetermined level of chance (i.e., at the alpha level set by the investigator).

At this point, an example may be helpful. Suppose someone handed you a penny and asked you to toss it 100 times and record the number of heads and tails. Assuming the coin was fair, a head or a tail should be equally likely to occur on any given toss. Therefore, in 100 tosses, your best guess about the outcome would be that 50 of the tosses will come out heads and 50 will come out tails. Your expected frequency distribution of heads and tails for 100 tosses of the coin is 50/50. It is this expected frequency against which your actual or obtained frequency can be compared. Most likely, you would be interested in whether the coin you had been handed was a fair, unbiased one. Following the basic logic of null hypothesis testing, you might set up the problem as one of testing $H_0$: The obtained frequency of heads and tails does not differ from the expected frequency. If in fact your obtained distribution was exactly 50/50, you could obviously not reject the null hypothesis and no statistical test would be needed. If your obtained distribution was 49/51 or even 48/52, in either direction, you would also probably not bother with a statistical test because it seems clear that such small deviations would occur often, even with a fair coin. But suppose you obtained 40 heads and 60 tails? This is not an intuitively clear-cut case, and a statistical test would be quite handy. Such a test would allow you to compute the likelihood of obtaining a split of 40/60 by chance when the true population frequency is 50/50. You could then set an alpha level, conduct the test, and draw an appropriate inference. In sum, then, all tests of frequency data require an expected frequency, an obtained frequency, and a means of determining the likelihood of obtaining various deviations from the expected frequency by chance. It should also be obvious that sample size will be as important here as it is with other tests. A fair coin is much more likely to show a 6/4 split by chance in 10 trials than it is to show a 60/40 split by chance in 100 trials, and therefore a larger sample will make it easier to detect small *real* differences than a smaller sample will. This is especially important because we cannot logically accept the null hypothesis. An obtained 6/4 split does not prove that the coin is not biased; it only means that we have failed to show that the coin *is* biased (which it may or may not be).

The logic described here applies to all statistical tests of frequency data, but other particulars of the tests vary according to the nature of the investigation. We will describe three of the most commonly used tests: the *binomial test*, the *chi-square test*, and the *Fisher exact probability test.*

### Binomial Test

In a literal sense, *binomial* means "two names," and so it should not be surprising that the binomial test deals with frequency data that fall into two nominal categories or classes. The number of situations involving two categories is large and includes Democrat-Republican, married-single, passed-failed, and alive-dead.

Like all statistical tests, the binomial test is based on a sampling distribution; the binomial distribution is the sampling distribution of the relative frequencies or proportions we would observe in successively drawn random samples of a given size from the same population. For example, suppose we tossed a fair coin three times and designated the number of heads obtained by the letter $f$. What is the probability that we would obtain each of the following outcomes: $f = 0, 1, 2,$ or 3?

The first column in Table 12-1a lists each of these outcomes; the second column enumerates the number of ways each could be obtained. For example, there are three ways to obtain $f = 2$: The first and second tosses could turn up heads (HHT), the first and third tosses could turn up heads (HTH), or the second and third tosses could turn up heads (THH). As Table 12-1a indicates, there are eight possible different sequences when a coin is tossed three times. The third column gives the probability of each sequence yielding a particular value of $f$. For $f = 2$, the probability of the sequence

$$\text{HHT} = (\tfrac{1}{2})(\tfrac{1}{2})(\tfrac{1}{2}) = .125$$

because the probability of a head $(P) = \frac{1}{2}$ and the probability of a tail $(Q) = \frac{1}{2}$ if the coin is fair. The fourth column gives the total probability of obtaining a particular value of $f$. For example, because the probability of each sequence containing two heads is equal to .125 and there are three such sequences, the probability that $f = 2 = 3(.125) = .375$. As you can see at the bottom of

TABLE 12-1 Possible outcomes of two binomial experiments of coin tossing, using a fair and an unfair coin

(a) $P = 1/2;\ Q = 1/2;\ N = 3$

| *Number of Heads (f)* | *Number of Ways* | | *Probability of Each Sequence* | *Probability f = Specified Value* |
|---|---|---|---|---|
| 0 | *TTT* | (1) | .125 | .125 |
| 1 | *HTT, THT, TTH* | (3) | .125 | .375 |
| 2 | *HHT, HTH, THH* | (3) | .125 | .375 |
| 3 | *HHH* | (1) | .125 | .125 |
| Total | | 8 | | 1.00 |

(b) $P = 1/3;\ Q = 2/3;\ N = 3$

| *Number of Heads (f)* | *Number of Ways* | | *Probability of Each Sequence* | *Probability f = Specified Value* |
|---|---|---|---|---|
| 0 | *TTT* | (1) | .295 | .295 |
| 1 | *HTT, THT, TTH* | (3) | .148 | .444 |
| 2 | *HHT, HTH, THH* | (3) | .074 | .222 |
| 3 | *HHH* | (1) | .037 | .037 |
| Total | | 8 | | 1.00 |

the last column, the combined probabilities of $f = 0, 1, 2,$ or $3 = 1.00$.

Figure 12-1a represents the theoretical sampling distribution of $f$ when $P = \frac{1}{2}$ and $N$ (number of tosses) = 3. The horizontal axis lists all possible values of $f$ (in this case, the number of heads), and the vertical axis gives a corresponding statement of the expected probability of each outcome.

Suppose, instead, that the coin you tossed was not fair and that the probability of a head was only $\frac{1}{3}$. An enumeration of the possible outcomes and their expected probabilities is given in Table 12-1b, and the sampling distribution is shown in Figure 12-1b. As you can see, the probabilities of outcomes including heads are now smaller. Specifically, the probability that $f = 2$ is found by multiplying 3 (the number of ways to get $f = 2$, or the number of sequences including 2 heads) by .074 [the probability of each of the sequences including 2 heads, e.g., $p(HTH) = (\frac{1}{3})(\frac{2}{3})(\frac{1}{3})$, or .074]. Therefore,

$$p(f = 2) = 3(.074) = .222.$$

As $N$ is increased, the process of actually enumerating all possible outcomes becomes more time consuming. However, we are able to apply a formula to obtain the information we need. More specifically, the binomial formula gives the probability of making $f$ observations in one category (e.g., heads) and $N - f$ observations in the other (e.g., tails). It is written

$$p(f) = \binom{N}{f}^{P^f Q^{N-f}}$$

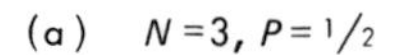


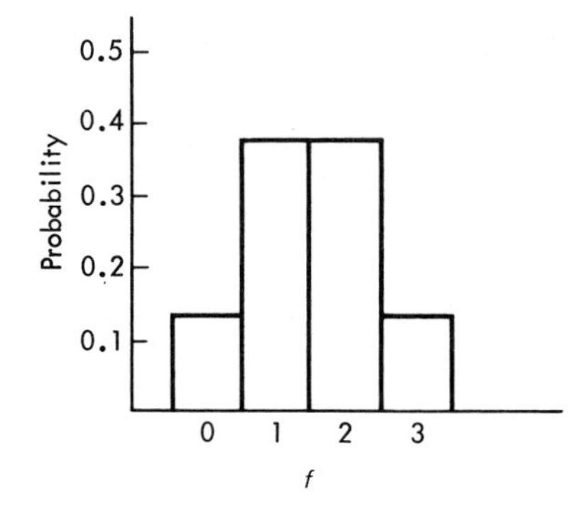


(b) $N = 3$, $P = 1/3$

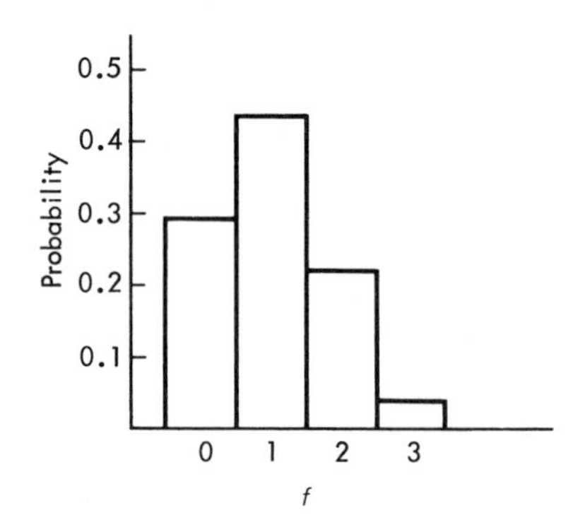


FIG. 12–1 Sampling distributions for the data in Table 12-1.

where $P$ = *expected* proportion of observations in one category

$Q = 1 - P$ = expected proportion of cases in the other category

$N$ = number of observations in the sample

$\binom{N}{f} = \dfrac{N!}{f!(N-f)!}$ = the so-called *binomial coefficient*

$N!$ = the mathematical expression (read as "$N$ factorial") that means $N(N-1)(N-2)\cdots(2)(1)$. Thus, when $N = 6$, $N! = (6)(5)(4)(3)(2)(1) = 720$, and when $N = 3$, $N! = (3)(2)(1) = 6$.

This formula has two basic parts: $P^f Q^{N-f}$ is the probability of obtaining any particular sequence containing $f$ observations in one category and $N - f$ in the other; and $\binom{N}{f}$ is the number of such sequences. To practice using this formula, verify the information in Table 12-1. For example, what is the probability of obtaining one head when a fair coin is tossed three times?

$p(f = 1)$ when $N = 3$ and $P = \frac{1}{2}$

$$p(f) = \frac{N!}{f!(N-f)!} P^f Q^{N-f}$$

$$p(f = 1) = \frac{3!}{1!2!}(\tfrac{1}{2})^1(\tfrac{1}{2})^2$$

$$= \frac{3 \cdot 2 \cdot 1}{1(2 \cdot 1)}(\tfrac{1}{2})^3$$

$$= 3(.125)$$

$$= .375$$

Although the binomial formula is quite complicated, especially with large $N$, its application is made considerably easier by the fact that there are tables which give the value of the binomial coefficient (by far the hardest part of the calculation) for various values of $N$. Appendix G shows the binomial coefficients for values of $N$ through 20.

Now that you know how to obtain the binomial sampling distribution when $P$ is known, let us see how this information can help when $P$ is not known and you want to test a hypothesis. Suppose we asked you to play a gambling game with us, involving the toss of a coin. If you were cautious, you might want to decide whether or not to play on the basis of whether you thought the coin we were using was fair. You might run an experiment by tossing the coin 10 times and recording the number of heads. Would you play if you observed 2 heads? 8 heads? 7 heads? Obviously, any outcome, even 0 or 10 heads, has some probability of occurring when the coin is fair. We will assume that you would want to be reasonably conservative about accusing us of cheating and therefore that you would not reject the hypothesis that $P = \frac{1}{2}$ unless you obtained an outcome that fell in the extreme .05 of the sampling distribution. Which values of $f$ would this decision rule include?

$$p(f = 0) = \frac{10!}{0!(10-0)!}(\tfrac{1}{2})^0(\tfrac{1}{2})^{10}$$

$$= 1(.00098)$$

$$= .0010$$

$$p(f = 1) = \frac{10!}{1!(10-1)!}(\tfrac{1}{2})^1(\tfrac{1}{2})^9$$

$$= 10(.00098)$$

$$= .0098$$

$$p(f = 2) = \frac{10!}{2!(10-2)!}(\tfrac{1}{2})^2(\tfrac{1}{2})^8$$

$$= 45(.00098)$$

$$= .0441$$

Because the probability distribution is symmetrical when $P = \frac{1}{2}$, the outcomes below would have the following probabilities:

$$p(f = 10) = .0010$$
$$p(f = 9) = .0098$$
$$p(f = 8) = .0441$$

The probability of obtaining an outcome as unusual as 0 or 10 heads is equal to $p(f = 0) + p(f = 10) = .0020$. The probability of obtaining an outcome as unusual as 0 or 1 head or 10 or 9 heads equals

$$p(f = 0) + p(f = 1) + p(f = 10) + p(f = 9) = .0216.$$

The probability of obtaining an outcome as unusual as 2 or fewer heads or 8 or more heads equals .1098. Therefore, to limit the probability of rejecting the hypothesis that $P = \frac{1}{2}$ when it is true (i.e., to limit the probability of falsely accusing us of cheating) to no more than .05, you should refuse to play if the number of heads you observe in 10 tosses is equal to 0, 1, 9, or 10.

*Normal approximation to the binomial distribution.* The procedure of calculating the exact probabilities of various outcomes takes longer as $N$ increases. However, we can take advantage of the fact that as $N$ increases, the binomial distribution looks more like a normal distribution. As an example, let us work out the previous problem using the normal approximation to the binomial. In Figure 12-2, a normal curve has been superimposed over the exact probabilities for various outcomes of our coin-toss example.

The mean of the binomial sampling distribution is equal to $NP$, and the standard deviation is equal to $\sqrt{NPQ}$. With this information, we can convert a particular value of $f$ to a standard score by the usual formula

$$z = \frac{\text{score} - \text{mean}}{\text{standard deviation}}$$

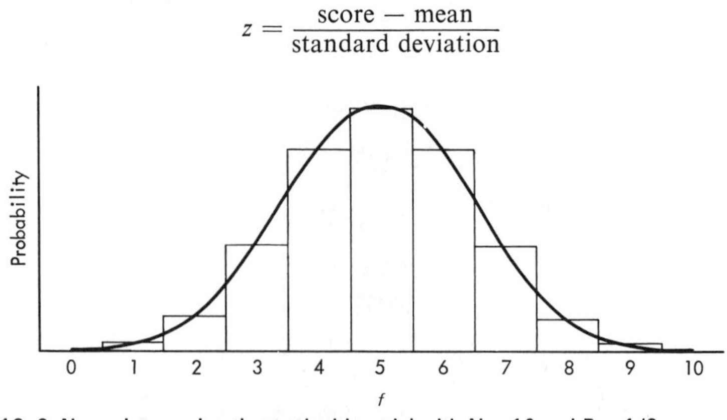


FIG. 12–2 Normal approximation to the binomial with $N = 10$ and $P = 1/2$.

or in this case

$$z = \frac{f - NP}{\sqrt{NPQ}}$$

Now, what is the standard score that corresponds to an outcome of 8 heads in our coin-toss experiment?

$$\begin{aligned} z &= \frac{8 - 10(\frac{1}{2})}{\sqrt{10(\frac{1}{2})(\frac{1}{2})}} \\ &= \frac{8 - 5}{1.58} \\ &= 1.90 \end{aligned}$$

According to the normal-curve table (Appendix C), .4713 of the area under the curve falls between the mean and a standard score of 1.90. Therefore, .0287 of the area falls above a score of 8. Our estimate of the area above 8 according to the exact probabilities we calculated earlier would be

$$.0441 + .0098 + .0010 = .0549.$$

As you can see, the estimate using the normal approximation to the binomial is a bit smaller and would lead us to conclude that a score as extreme as 8 is less probable than it really is. Our probability estimate can be made more accurate by using the corrected standard score formula given below:

$$z = \frac{|f - NP| - 0.5}{\sqrt{NPQ}}$$

As you can see, 0.5 is subtracted from the absolute value of the difference between our obtained and our expected scores. The effect of this is to reduce the size of the standard score for any particular value of $f$. The corrected standard score for the present example is

$$\begin{aligned} z &= \frac{|8 - 5| - 0.5}{\sqrt{10(\frac{1}{2})(\frac{1}{2})}} \\ &= \frac{3 - 0.5}{1.58} \\ &= \frac{2.5}{1.58} \\ &= 1.58 \end{aligned}$$

According to Appendix C, the area between the mean and 1.58 is equal to .4429, and therefore the probability of scores of 8 or larger equals

$$.5000 - .4429 = .0571.$$

This compares very well with the value .0549 calculated previously. The correction is generally applied to improve probability estimates when the normal approximation to the binomial is used. It is necessary because we are using a continuous distribution to estimate the probabilities under a discrete distribution. The normal approximation assumes that values of $f$ such as 3.65 can occur; whereas you can in fact have 3 or 4 heads, but you can never have 3.65 heads. This correction has less effect as $N$ gets larger, as you can see by inspecting the formula. This makes sense because, as we said in the beginning of this section, as $N$ increases, the binomial distribution is more like the normal distribution. Furthermore, the closer $P$ and $Q$ are to $\frac{1}{2}$, the better the match between the normal and the binomial for any given value of $N$. You can see why this would be the case by referring back to Figure 12-1. The sampling distribution when $P = \frac{1}{2}$ is symmetrical, whereas the sampling distribution when $P = \frac{1}{3}$ is not. Because the normal distribution is symmetrical, values of $P$ that deviate from $\frac{1}{2}$ will necessarily produce some error when the normal approximation to the binomial is used. In general, when $P$ is very close to $\frac{1}{2}$ and $N$ is fairly large (e.g., 20), the normal approximation to the binomial is a handy substitute for the more laborious procedure of calculating probabilities of binomial outcomes.

EXAMPLE 1

Suppose a person came to us and claimed to have the power of psychokinesis (controlling events through the power of the mind) and maintained that he could make a fair die come up "one" at least some of the time by concentrating on it as he rolled. How could we test this claim? One way would be to let the subject roll a die 12 times, for example, and observe the number of times "one" occurred. Suppose "one" occurred on 5 of the 12 trials. Would this suggest that the subject can control the outcome of the die? The first question we would ask ourselves is how many times would we expect "one" to come up if the subject could not affect the roll. Because the probability of "one" occurring with a fair die is $\frac{1}{6}$, we would expect "one" $\frac{1}{6}$ of the time, or on 2 of the 12 trials. Because we observed over twice as many "ones" as this, the evidence for our subject's claim looks pretty good. However, we also know that we could have observed as many as 12 "ones" by chance, without any help from the subject at all. Therefore, we need to determine the likelihood of observing 5 or more "ones" in 12 trials by chance. This is equal to the sums of the probabilities of the obtained outcome, and each of the more extreme outcomes that would also have supported our subject's claim.

In other words

$$p(f \geq 5) = p(5) + p(6) = p(7) + p(8) + p(9) + p(10) + p(11) + p(12)$$

where $N = 12$ (number of observations)

$P = \frac{1}{6}$ (expected proportion of "ones" when rolling a fair die)

$Q = \frac{5}{6}$ (expected proportion of outcomes other than "one," i.e., 2, 3, 4, 5, or 6)

This is a case in which we would want to compute exact probabilities because $N$ is relatively small and, especially, because $P$ is not close to $\frac{1}{2}$. The relevant calculations, which you can do yourself as a check on your understanding, show that in this case

$$p(5) = 0.028;\ p(6) = 0.007;\ p(7) = 0.001;\ p(8) = 0.0001;\ p(9) = 0.000;$$
$$p(10) = 0.000,\ p(11) = 0.000;\ p(12) = 0.000.$$

Therefore

$$p(f \geq 5) = 0.028 + 0.007 + 0.001 + 0.000 + 0.000 + 0.000 + 0.000 + 0.000^{*} = 0.036$$

With a .05 alpha level, we could conclude that our subject has special powers of psychokinesis.

EXAMPLE 2

In order to determine taste preferences in a certain species of rat, 25 animals were given free and simultaneous access to chocolate milk and orange juice. The rats drank from containers that were calibrated to indicate the amount of liquid consumed. The results are given in Table 12-2.

In general, do the rats prefer one or the other liquid? To answer this question, you could perform a within-subject $t$ test (see Chapter 7) on the difference scores listed in the third column of Table 12-2. You could also use the techniques discussed in this chapter. If you assume that the animals do not have any preference, the probability of any particular animal drinking more chocolate milk than orange juice should be $\frac{1}{2}$. In the fourth column of Table 12-2, animals that drank more chocolate milk are assigned a plus sign, and animals that drank more orange juice are assigned a minus sign (hence, this test is sometimes called the *sign test*). If $P(+) = \frac{1}{2}$, what is the probability of obtaining an outcome as extreme as 16 +'s in 25 trials?

*These last five values are slightly above 0, of course, but they are so unlikely to occur by chance that they are inconsequential.

TABLE 12-2 Data from an experiment on taste preferences in rats

| *Rat* | *Chocolate Milk* | *Orange Juice* | *D* | *Sign of the Difference* |
|---|---|---|---|---|
| 1 | 7 | 3 | 4 | + |
| 2 | 9 | 6 | 3 | + |
| 3 | 12 | 7 | 5 | + |
| 4 | 1 | 3 | −2 | − |
| 5 | 10 | 2 | 8 | + |
| 6 | 5 | 1 | 4 | + |
| 7 | 6 | 7 | −1 | − |
| 8 | 12 | 7 | 5 | + |
| 9 | 20 | 22 | −2 | − |
| 10 | 16 | 12 | 4 | + |
| 11 | 23 | 19 | 4 | + |
| 12 | 14 | 17 | −3 | − |
| 13 | 11 | 5 | 6 | + |
| 14 | 9 | 14 | −5 | − |
| 15 | 20 | 14 | 6 | + |
| 16 | 12 | 13 | −1 | − |
| 17 | 17 | 12 | 5 | + |
| 18 | 10 | 4 | 6 | + |
| 19 | 6 | 2 | 4 | + |
| 20 | 7 | 9 | −2 | − |
| 21 | 14 | 8 | 6 | + |
| 22 | 9 | 10 | −1 | − |
| 23 | 6 | 1 | 5 | + |
| 24 | 27 | 22 | 5 | + |
| 25 | 5 | 7 | −2 | − |

Because $N = 25$ and $P = \frac{1}{2}$, this is a case in which the normal approximation to the binomial would be appropriate.

$$z = \frac{|f - NP| - 0.5}{\sqrt{NPQ}}$$

$$= \frac{|16 - 25(\frac{1}{2})| - 0.5}{\sqrt{25(\frac{1}{2})(\frac{1}{2})}}$$

$$= \frac{3}{2.5}$$

$$= 1.20$$

According to Appendix C, the area between the mean and a standard score of 1.20 is .3849, and therefore the area above 1.20 is .1151. With an alpha level of .05, only standard scores falling in the upper (indicating a preference for chocolate milk) or the lower (indicating a preference for orange juice) .025 of the curve would lead us to reject the hypothesis that $P(+) = \frac{1}{2}$. Therefore, we do not have reason to conclude that this species of rat prefers one or the other liquid.

### Chi-square Test

A great deal of social science research involves the comparison of two or more independently drawn samples to determine the likelihood that they have been drawn from equivalent populations. When the data involved in such a study fall into discrete categories, the investigator will obtain a set of frequencies for each sample and the *chi-square* ($\chi^2$) *test* can be used. We will illustrate its use in an example and explain the logic of the test as we go along.

Suppose we wished to determine whether children are more likely to share candy with another child if they have just seen an example of sharing than if they have not. Using the experimental method, we randomly assign 80 children to either the Treatment group (those who see a sharing model before being tested) or the Control group (those who see no model). All children are then put into a situation in which they must choose between sharing and not sharing with a friend, and their reactions are recorded. Let us suppose in this example that there were 40 children in the Treatment group and 40 in the Control group. In addition, we will suppose that 25 of the Treatment group children shared but that only 10 of the Control group children did so. Our problem is to find out how likely this difference in frequency would be to occur by chance, that is, if the example of sharing really had no influence. We must compare the frequencies 25/15 and 10/30 with a statistical test.

The test we require will have to be based on the sampling distribution that would be obtained if there were no real differences in the populations (i.e., if the treatment had no effect) and if we repeated the experiment many times. We do not actually have to carry out the process, of course, because the relevant distribution is already available to us in the form of the sampling distribution for the statistical test, $\chi^2$. To make use of it, we must set up our data in a particular way.

Table 12-3 shows the proper setup for our hypothetical example. The study's outcome really provides us with four cells of information, each of which is a frequency. Specifically, 25 children who saw a sharing model also shared themselves, 15 children who saw a sharing model did not share, 10 children who did not see a sharing model shared anyhow, and 30 children who did not see a sharing model also did not share. These four frequencies form a $2 \times 2$ table containing the obtained outcome. Also note in Table 12-3 that the marginal totals provide combined information about the proportion of children who did and did not share overall (the column totals) and the proportion who were and were not given the treatment (the row totals).

The cells of the table, then, provide the obtained outcome needed for the statistical test. But what of our expected outcomes? They are estimated from the marginal totals because it can be shown that the data within the cells of the table would be proportional to the marginal totals if the treatment made no difference. The difference between the obtained and expected proportions will be tested for statistical significance according to much the same reason-

TABLE 12-3 Outcome of an experiment concerning the effects of a sharing model on children's sharing. The dependent measure is whether or not the child shared, expressed according to whether the child was in the Treatment or the Control condition

(a) Basic Data in Table Form

| | *Shared* | *Did Not Share* | *Total* |
|---|---|---|---|
| Exposed to model | 25 | 15 | 40 |
| Control (no model) | 10 | 30 | 40 |
| Total | 35 | 45 | 80 |

(b) Expected (Italic) and Observed Frequencies

| | *Shared* | *Did Not Share* | *Total* |
|---|---|---|---|
| Exposed to model | *17.5* 25 | *22.5* 15 | 40 |
| Control (no model) | *17.5* 10 | *22.5* 30 | 40 |
| Total | 35 | 45 | 80 |

ing that underlies the binomial test, but the actual computational procedures are of course different.

The first step is to compute the expected outcome for each cell from the marginals. This is done by multiplying the two marginal totals common to a cell and dividing the product by the overall total, $N$. Table 12-3b shows that the expected value in the cell for children who saw a sharing model and subsequently shared is 17.5. The value was obtained by multiplying the number of children exposed to a model (the marginal total, 40) by the number of children who shared (the marginal total, 35) and dividing the product by the total number of children in the experiment (80). Corresponding calculations were used to compute the other three expected values, as you can check for yourself.

We can now compute the statistic, $\chi^2$. Its formula is

$$\chi^2 = \sum \frac{(O - E)^2}{E}$$

The formula directs us to subtract each expected value ($E$) from its corresponding obtained value ($O$) and to square this difference. The squared difference is then divided by the cell's $E$. This procedure is followed for each cell. We then sum the resulting values over all columns and rows (i.e., across all four cells), and this sum gives us the appropriate value of $\chi^2$. Thus, in our example

$$\chi^2 = \frac{(25 - 17.5)^2}{17.5} + \frac{(15 - 22.5)^2}{22.5} + \frac{(10 - 17.5)^2}{17.5} + \frac{(30 - 22.5)^2}{22.5}$$

$$= \frac{7.5^2}{17.5} + \frac{-7.5^2}{22.5} + \frac{-7.5^2}{17.5} + \frac{7.5^2}{22.5}$$

$$= \frac{56.25}{17.5} + \frac{56.25}{22.5} + \frac{56.25}{17.5} + \frac{56.25}{22.5}$$

$$= 3.21 + 2.5 + 3.21 + 2.5$$

$$= 11.42$$

The probability that the proportions shown in the table depart from chance can then be determined by using the table in Appendix *H*, but this requires finding the appropriate *df* first. The *df* for such a table, usually referred to as a *contingency table*, is given by the formula

$$df = (r - 1)(k - 1)$$

where $r$ = number of rows in the table

$k$ = number of columns in the table

Therefore, in the present case, $df = (2 - 1)(2 - 1) = 1$

Suppose we had set $\alpha = .05$. We would then look up the critical value of $\chi^2$ in Appendix *H* for $df = 1$. Inasmuch as our obtained value, 11.42, exceeds the critical value of 3.84, we can reject the null hypothesis and conclude that the observed proportion of sharers to nonsharers differs, or is contingent upon whether the children were exposed to a sharing model or not. Children who saw the model were more likely to share than children who did not see the model.

*A further example of the use of* $\chi^2$. In the preceding example, an equal number of children were assigned to the cells of a simple experiment, resulting in a $2 \times 2$ contingency table. But $\chi^2$ can be used in correlational studies as well as experiments, equal $n$'s are not required in each cell, and the contingency tables can be of any size. The same basic computational formula still applies.

Suppose an investigator wanted to know whether the proportion of children who participate in organized extracurricular sports changes as children grow older. All the children in the first three grades of a local elementary school are surveyed, and a determination is made about whether each child does or does not participate in after-school athletics. Table 12-4 shows the resulting contingency table, in which the expected values have been calculated according to the procedure already outlined. Now, following the formula

$$\chi^2 = \sum \frac{(O - E)^2}{E}$$

$$= \frac{(12 - 11.15)^2}{11.5} + \frac{(17 - 17.85)^2}{17.85} + \frac{(18 - 18.46)^2}{18.46} + \frac{(30 - 29.54)}{29.54}$$

$$+ \frac{(15 - 15.38)^2}{15.38} + \frac{(25 - 24.62)}{24.62}$$

$$= \frac{0.85^2}{11.5} + \frac{-0.85^2}{17.85} + \frac{-0.46^2}{18.46} + \frac{0.46^2}{29.54} + \frac{-0.38^2}{15.38} + \frac{0.38^2}{24.62}$$

$$= \frac{0.72}{11.5} + \frac{0.72}{17.85} + \frac{0.21}{18.46} + \frac{0.21}{29.54} + \frac{0.14}{15.38} + \frac{0.14}{24.62}$$

$$= 0.06 + 0.04 + 0.01 + 0.01 + 0.01 + 0.01$$

$$= 0.14$$

The appropriate *df* for this $\chi^2$ is $(3 - 1)(2 - 1)$ because the table has three rows and two columns. Thus, assuming we have again set $\alpha$ at .05, we must consult Appendix H and find the critical value of $\chi^2$ with $df = 2$ and $\alpha = .05$. It is 5.99, which clearly is not exceeded by our obtained $\chi^2$ of .14. The null hypothesis therefore cannot be rejected. We cannot conclude that the proportion of children participating in after-school athletics is contingent upon their grade in school.

One final point should be noted: The first step in computing any $\chi^2$ is to set up the data in the form of a contingency table. When doing so, you always have to decide which variable will be designated in the rows and which in the columns. However, regardless of what your decision is, the formula will give exactly the same value of $\chi^2$ in either case. Verify this for yourself by using the setup in Table 12-4b to compute the expected frequencies and $\chi^2$. Your values should correspond exactly to those just computed for Table 12-4a.

*Correction for continuity.* When $df = 1$, when the total $N$ is less than 40, or when the *expected* value of any one cell is less than 10, our original $\chi^2$ formula is not appropriate without a so-called *correction for continuity.* The adjustment has a fairly complicated mathematical rationale, but like the correction applied when the normal approximation to the binomial is used, it boils down to the fact that the correction makes a better fit between the computed value of $\chi^2$ in these cases and the underlying $\chi^2$ distribution on which the test is based. The corrected formula is

$$\chi^2 = \sum \frac{(|O - E| - .5)^2}{E}$$

TABLE 12-4 Obtained and expected values for a correlational study of the proportion of children at three grade levels who do and do not participate in after-school athletic activities

(a) Expected (Italic) and Observed Frequencies

| | *Athletics* (Expected) | Athletics (Observed) | *No Athletics* (Expected) | No Athletics (Observed) | *Total* |
|---|---|---|---|---|---|
| First grade | *11.15* | 12 | *17.85* | 17 | 29 |
| Second grade | *18.46* | 18 | *29.54* | 30 | 48 |
| Third grade | *15.38* | 15 | *24.62* | 25 | 40 |
| Total | | 45 | | 72 | 117 |

(b) An Alternative Arrangement of the Data

| | *First* | *Second* | *Third* |
|---|---|---|---|
| Athletics | 12 | 18 | 15 |
| No athletics | 17 | 30 | 25 |

An example will help to demonstrate the application of this formula. Suppose there are 28 people of voting age in a certain rural community, all of whom are registered as either Democrats or Republicans. You are able to find out which of these people voted in the last local election (dominated by Republicans), and you want to know whether the likelihood of a person's voting is contingent on the party to which he or she belongs. Fifteen of the citizens are Republicans, and 10 of them voted; the remaining 13 people are registered Democrats, and 7 of them voted. The data are presented in Table 12-5, along with the expected cell frequencies (for practice, you should verify these values).

Now, applying the correction

$$\chi^2 = \frac{(|7 - 7.89| - 0.5)^2}{7.89} + \frac{(|10 - 9.11| - 0.5)^2}{9.11} + \frac{(|6 - 5.11| - 0.5)^2}{5.11} + \frac{(|5 - 5.89| - 0.5)^2}{5.89}$$

$$= \frac{0.39^2}{7.89} + \frac{0.39^2}{9.11} + \frac{0.39^2}{5.11} + \frac{0.39^2}{5.89}$$

$$= 0.019 + 0.017 + 0.030 + 0.026$$

$$= 0.092$$

TABLE 12-5 Proportion of registered Democrats and Republicans who did and did not vote in a local election, requiring the use of the $\chi^2$ formula corrected for continuity

| | *Democrat* | *Republican* | *Total* |
|---|---|---|---|
| Voted | *7.89* 7 | *9.11* 10 | 17 |
| Did not vote | *5.11* 6 | *5.89* 5 | 11 |
| Total | 13 | 15 | 28 |

This is a 2 × 2 contingency table, and therefore because $r = 2$, $k = 2$, and $df = (r - 1)(k - 1)$, we will be using the $\chi^2$ distribution in Appendix H for $df = 1$. The critical value when $\alpha$ is set at .05 is 3.84, which is far greater than our obtained value, and we therefore cannot conclude that voting was related to party affiliation.

### Fisher Exact Probability Test

We have said that $\chi^2$ must be corrected when $N$ is less than 40 or when the expected value in any one cell is less than 10. Unfortunately, when the expected value in any one cell is less than 5, $\chi^2$ becomes useless; that is, the computed statistic cannot be interpreted with the $\chi^2$ distribution. In these small $N$ cases, involving 2 × 2 contingency tables, the Fisher exact probability test may be used. As the name implies, Fisher's test provides the exact probability that the 2 × 2 contingency table will take a particular form, according to the formula

$$p = \frac{(A + B)!(C + D)!(A + C)!(B + D)!}{N!A!B!C!D!}$$

The symbols $A$ to $D$ represent the obtained values in each of the cells of the table, as shown in Table 12-6. $N$ is, of course, the total number of observations. (Recall the factorial symbol, !, was defined on page 184.)

Suppose we had used a small sample of cars to compare radial and standard tires after 15,000 miles of driving to see whether they still had adequate tread left. The possible outcome shown in Table 12-6a lends itself readily to the Fisher test. Specifically, by substitution

$$p = \frac{7!5!5!7!}{12!0!7!5!0!}$$

$$= .0013$$

TABLE 12-6 Hypothetical outcomes of a study of the longevity of automobile tires as a function of whether they are of the radial or standard type. Fisher's exact probability test should be applied to these data

(a)

| | *Worn Out After 15,000 Miles* | *Not Worn Out* | *Total* |
|---|---|---|---|
| Radial | *A* 0 | *B* 7 | 7 |
| Standard | *C* 5 | *D* 0 | 5 |
| Total | 5 | 7 | 12 |

(b)

| | *Worn Out After 15,000 Miles* | *Not Worn Out* | *Total* |
|---|---|---|---|
| Radial | *A* 1 | *B* 6 | 7 |
| Standard | *C* 4 | *D* 1 | 5 |
| Total | 5 | 7 | 12 |

The exact probability of obtaining this particular contingency outcome is .0013. We would surely conclude, if the data were real, that radial tires do have a longer life-span than standard tires.

Now suppose that our obtained result had been the one shown in Table 12-6b. We could apply the Fisher formula to these values, but that would give us only the likelihood of obtaining the one outcome shown in Table 12-6b. However, our real question is: What is the likelihood of obtaining this particular outcome *or one that supports the experimental hypothesis even more strongly*? We did not encounter this issue in analyzing the data in Table 12-6a because with these particular marginals we already had the most extreme difference possible (i.e., all the standard tires had worn out, and none of the radials had).

To analyze the data in Table 12-6b correctly using the Fisher exact probability test, we would first compute the exact probability of the obtained outcome:

$$p = \frac{7!5!5!7!}{12!1!6!4!1!}$$

$$= .044$$

Next, we would compute the exact probabilities of each of the more extreme outcomes that might have occurred. In this case, there is only one outcome that is more extreme, the one shown in Table 12-6a, and we have already determined that its exact probability is .0013. The final step is to sum the exact probabilities of the obtained outcome and all the more extreme outcomes. This sum is the true confidence level you would have in rejecting the null hypothesis. In our example, the exact probability of obtaining a sample outcome at least as extreme as the one shown in Table 12-6b is

$$p = .044 + .0013 = .045$$

This value is less than .05, and so if we had initially set $\alpha$ at .05, we would be able to reject the null hypothesis and infer that radial tires are more long-lived than standard tires.

## DEALING WITH RANKS

### Mann-Whitney *U* Test

This test is a useful technique when you want to compare two samples and only the rank order of the scores is available or the scores have been converted to ranks. It is an alternative to the independent-groups $t$ test; however, in this case, our null hypothesis is that the two samples come from populations with the same distribution, rather than (as in the case of $t$) from two populations with the same mean. For example, Table 12-7a lists hypothetical achieve-

TABLE 12-7 Hypothetical scores showing the effects of training on achievement

(a)

| *Training Program* | | *No Training Program* | |
|---|---|---|---|
| *Score* | *Rank* | *Score* | *Rank* |
| 30 | (4) | 36 | (6) |
| 17 | (1) | 31 | (5) |
| 60 | (12) | 43 | (8) |
| 47 | (10) | 23 | (2) |
| 50 | (11) | 27 | (3) |
| 40 | (7) | 46 | (9) |
| 244 | | 206 | |

(b)

| Rank | 1 | 2 | 3 | 4 | 5 | 6 | 7 | 8 | 9 | 10 | 11 | 12 |
|---|---|---|---|---|---|---|---|---|---|---|---|---|
| Score | 17 | 23 | 27 | 30 | 31 | 36 | 40 | 43 | 46 | 47 | 50 | 60 |
| Condition | T | N | N | T | N | N | T | N | N | T | T | T |

ment scores for an experiment in which a random half of the participants received a brief training program prior to the test and the other half did not.

The numbers in parentheses beside each score indicate the relative standing of each person in the study. That is, the person with the lowest score received a rank of 1, the person with the second lowest score received a rank of 2, and the person with the highest score received a rank of 12. An inspection of Table 12-7a suggests that the people who participated in the training program tend to have higher ranks than those who did not. This is easier to see in Table 12-7b, in which the scores have been arranged in order and the condition corresponding to each score is noted.

Now count the number of No-Training people who did better than each Training person. Six No-Training people (those with scores of 23, 27, 31, 36, 43, and 46) did better than the first Training subject (the one whose score was 17). Four No-Training subjects (those with scores of 31, 36, 43, and 46) did better than the next Training subject (the one whose score was 30). Two (those with scores of 43 and 46) did better than the next Training subject (score, 40) and 0 No-Training subjects did better than each of the last three Training subjects. The sum of the number of No-Training people who did better than each Training person equals $6 + 4 + 2 + 0 + 0 + 0 = 12$. Let this value equal $U_1$. Now let $U_2$ equal the number of Training people who did better than each No-Training person: $U_2 = 5 + 5 + 4 + 4 + 3 + 3 = 24$. If, overall, there was no difference between these two conditions, $U_1$ should equal $U_2$ because, on the average, you would expect those who had training to precede those who had no training about as often as those with no training precede those with training. On the other hand, a very *small* value of $U_1$ (and a correspondingly large value of $U_2$) would mean that No-Training people rarely perform better than Training people, indicating an interesting difference between the two groups. Therefore, large differences in the values of $U_1$ and $U_2$ will lead us to reject the null hypothesis that the samples come from similar distributions. The sampling distribution of $U$ (which is traditionally defined as the smaller of $U_1$ or $U_2$) depends upon the size of each sample. Appendix I gives the critical values of $U$ for samples of various sizes. For $n_1 = 6$ and $n_2 = 6$, $U$ can be *no greater than* 5 in order for us to reject the null hypothesis at the 0.05 level. (Note that this table is read somewhat differently from those presented previously.) Because $U_1$ (the smaller of our two obtained values) $= 12 > 5$, we cannot conclude that the training program affected performance in this case.

When $N$ is quite large, the above procedure for obtaining the values of $U_1$ and $U_2$ is much more time consuming. However, the following formulas yield the same results:

$$U_1 = n_1 n_2 + \frac{n_1(n_1 + 1)}{2} - \sum R_1$$

and

$$U_2 = n_1 n_2 + \frac{n_2(n_2 + 1)}{2} - \sum R_2$$

where $n_1$ = size of the first sample

$R_1$ = sum of the ranks assigned to the cases in the first sample

$n_2$ = size of the second sample

$R_2$ = sum of the ranks assigned to the cases in the second sample

As a check on your calculations, $U_1 + U_2$ should equal $(n_1)(n_2)$.

The data in Table 12-8 represent the number of social and business contacts in a day for a hypothetical sample of 5 people living in urban environments and 4 people living in rural environments. Ranks are given in parentheses, and the sum of the ranks in each group is given at the bottom of the appropriate column. Note that when two scores are tied, they are given the average of the next two ranks. In this example, the two people who should have ranks 2 and 3 both have the same score and hence are each assigned a rank of 2.5. The next person is assigned the rank of 4, and so on. The same reasoning applies when more people are tied; for example, if the three lowest scores are all the same, the three subjects would each receive a rank of 2 [(1 + 2 + 3)/3].

According to the formulas

$$\begin{aligned} U_1 &= n_1 n_2 + \frac{n_1(n_1 + 1)}{2} - \sum R_1 \\ &= (5)(4) + \frac{(5)(6)}{2} - 34 \\ &= 20 + 15 - 34 \\ &= 1 \end{aligned}$$

and

$$\begin{aligned} U_2 &= n_1 n_2 + \frac{n_2(n_2 + 1)}{2} - \sum R_2 \\ &= (5)(4) + \frac{(4)(5)}{2} - 11 \\ &= 20 + 10 - 11 \\ &= 19 \end{aligned}$$

According to Appendix I, $U$ can be 1 (but no larger) to be significant at the .05 level. Therefore, we may reject the null hypothesis and conclude that the

TABLE 12-8 Number of social and business contacts of urban and rural residents

| *Urban* | | *Rural* | |
|---|---|---|---|
| *Score* | *Rank* | *Score* | *Rank* |
| 4 | (4) | 5 | (5) |
| 9 | (7) | 0 | (1) |
| 10 | (8) | 3 | (2.5) |
| 12 | (9) | 3 | (2.5) |
| 7 | (6) | | |
| | 34 | | 11 |

distribution of the number of daily social and business contacts is not the same in the urban and rural populations from which these samples were drawn.

*Normal approximation for the Mann-Whitney U test.* As sample size increases, the normal curve can be used to approximate the sampling distribution of $U$. The mean of the sampling distribution of $U = n_1 n_2/2$, and the standard deviation $= \sqrt{n_1 n_2(n_1 + n_2 + 1)/12}$. Therefore, our obtained value of $U$ can be put in standard-score form according to the following formula

$$z = \frac{U - (n_1 n_2)/2}{\sqrt{n_1 n_2(n_1 + n_2 + 1)/12}}$$

This standard score is then evaluated in the usual fashion, using the normal-curve table in Appendix C. The normal approximation is generally not used unless at least one of the samples consists of more than twenty observations.

### Spearman's Rank-Difference Correlation

Spearman's rank-difference correlation (rho) is a measure of the association of two variables when the scores are in ranked form. For example, the data in the first two columns of Table 12-9 represent the hypothetical rankings of a sample of 10 nursing homes on two variables: quality of health care and size of resident population. The third column is the difference in ranks on the two variables for each home, and the fourth column contains the squares of these difference scores. You could compute the Pearson correlation coefficient (see Chapter 11) using these ranks as scores. Rho is based on the squared-difference scores, and is easier to use when the data are in the form of ranks and $N$ is relatively small.

$$\text{rho} = 1 - \frac{6(\sum D^2)}{n(n^2 - 1)}$$

TABLE 12-9 Study of the relationship between the quality of health care and the number of patients in nursing homes

| *Nursing Home* | *Quality of Health Care (rank)* | *Size of Patient Population (rank)* | *D* | *D*² |
|---|---|---|---|---|
| A | 3 | 8 | −5 | 25 |
| B | 8 | 4 | 4 | 16 |
| C | 1 | 10 | −9 | 81 |
| D | 7 | 5 | 2 | 4 |
| E | 4 | 9 | −5 | 25 |
| F | 2 | 7 | −5 | 25 |
| G | 6 | 6 | 0 | 0 |
| H | 5 | 1 | 4 | 16 |
| I | 9 | 2 | 7 | 49 |
| J | 10 | 3 | 7 | 49 |
| | | | 0 | 290 |

In the present example

$$\text{rho} = 1 - \frac{6(290)}{10(10^2 - 1)}$$
$$= 1 - \frac{1{,}740}{990}$$
$$= 1 - 1.76$$
$$= -.76$$

In order to determine whether this evidence should lead us to reject the hypothesis that the population value of rho = 0, consult Appendix J. With 10 observations, the absolute value of our obtained rho must exceed .65 in order to be significant at the .05 level. Because our value is −.76, we may conclude that there is a significant negative correlation between health care and the number of residents. Ties in ranks on each variable would be handled as they are in the Mann-Whitney U test.

## Practice Problems

### *A. For My Next Act*

A person claiming to have the power of clairvoyance (the power to know events in advance) watches you roll a fair die 10 times after predicting that you will roll 4 "twos." This is, in fact, the outcome. What is the probability of rolling exactly 4 "twos" in 10 rolls? Would you conclude that the person has clairvoyance?

### *B. An Ounce of Prevention*

As part of a rehabilitation program, half of a group of male ex-convicts, selected randomly, received counseling on how to make their marriages successful, and the other half received an equal amount of personal attention that did not involve marriage counseling. The number of men who were divorced five years later is given below:

| | Still Married | Divorced |
|---|---|---|
| Marriage counseling | 70 | 30 |
| No marriage counseling | 40 | 60 |

Was the counseling program effective?

### *C. Birds of a Feather*

Jane was trying to decide whether or not to end her relationship with Bill. Part of the problem, she thought, was that they did not agree on which characteristics she ought to have. She chose nine characteristics from a text left over from an old personality course and ordered them according to how important she thought it was for her to develop them. She also asked Bill to order them according to how important he thought it was for her to develop them. She made up her mind that if there was no evidence of agreement, she would find someone else.

| Jane's List | Bill's List |
|---|---|
| intelligent | affectionate |
| affectionate | loyal |
| loyal | cheerful |
| witty | modest |
| practical | practical |
| cheerful | intelligent |
| patient | patient |
| modest | witty |
| sophisticated | sophisticated |

What did Jane do?

### *D. What's in a Name?*

The number of people buying each of four brands of toothpaste at a local drugstore one day is given below:

| Twinkle | Toothglow | Fang | Smile-white |
|---|---|---|---|
| 16 | 7 | 5 | 20 |

Assuming that this sample is adequately representative, can you infer that the brands differ in their popularity?

## *E. Shedding Light on the Subject*

A lamp factory hired a research consulting firm to investigate the effectiveness of their assembly-line procedures. Specifically, they wondered if workers would be more productive if they changed from one assembly job to another (shades, fixtures, cords) during a day rather than remaining on a single job all day. Eight people were assigned to each of two conditions, and the number of units produced in three days was recorded for each subject.

| Three Different Jobs | Same Job |
|---|---|
| 120 | 65 |
| 99 | 500 |
| 100 | 15 |
| 80 | 60 |
| 25 | 75 |
| 98 | 20 |
| 90 | 24 |
| 89 | 23 |

Using a ranking method, determine whether the two groups differed.

## *F. Nobody Knows Noses?*

Actors were hired to express love or anger, and their faces were photographed. Masks placed on each photo allowed only the nose to show, and the photo's were given to subjects who were asked to rate each photo on a scale of 1 to 10, with 1 representing a definite expression of love and 10 a definite expression of anger. Average ratings for photos of each type for each of 36 subjects is given below:

| Subject | Photos portraying love | Photos portraying anger |
|---|---|---|
| 1 | 3.7 | 4.2 |
| 2 | 4.0 | 1.3 |
| 3 | 3.4 | 5.1 |
| 4 | 7.3 | 6.2 |
| 5 | 6.8 | 7.1 |
| 6 | 2.2 | 2.3 |
| 7 | 8.1 | 7.3 |
| 8 | 9.7 | 9.8 |
| 9 | 5.6 | 7.8 |
| 10 | 2.5 | 2.4 |
| 11 | 4.3 | 6.2 |
| 12 | 1.5 | 8.3 |
| 13 | 2.2 | 2.9 |
| 14 | 3.6 | 4.5 |
| 15 | 1.1 | 1.4 |
| 16 | 7.8 | 8.3 |

| | | |
|---|---|---|
| 17 | 7.2 | 6.3 |
| 18 | 5.1 | 5.0 |
| 19 | 4.8 | 4.3 |
| 20 | 6.2 | 6.5 |
| 21 | 3.6 | 4.2 |
| 22 | 2.1 | 3.0 |
| 23 | 4.5 | 4.7 |
| 24 | 5.0 | 4.2 |
| 25 | 6.8 | 7.2 |
| 26 | 3.3 | 3.1 |
| 27 | 9.3 | 9.8 |
| 28 | 6.2 | 3.6 |
| 29 | 4.5 | 4.0 |
| 30 | 8.3 | 7.6 |
| 31 | 4.6 | 5.2 |
| 32 | 5.3 | 7.2 |
| 33 | 6.0 | 5.7 |
| 34 | 6.1 | 6.3 |
| 35 | 2.2 | 2.7 |
| 36 | 2.6 | 2.8 |

Could the subjects tell the difference between the two types of photographs?

## *G. The Numbers Game*

Look again at Problem E. Compute the mean number of units produced in each condition. Because the mean performance of the Same-Job subjects is greater than the mean performance of the Three-Different-Jobs subjects, how could the results of your Mann-Whitney $U$ test indicate that the Three-Different-Jobs condition is superior?

# Glossary of Formulas

| Formula | *Definition* |
| --- | --- |
| $a = \bar{Y} - b(\bar{X})$ | *Y–intercept of linear regression equation* |
| $b = \frac{SS_{XY}}{SS_X} = \frac{\sum (X - \bar{X})(Y - \bar{Y})}{\sum (X - \bar{X})^2}$ | *Defining formula for slope of the linear regression line* |
| $b = \frac{SS_{XY}}{SS_X} = \frac{\sum XY - \frac{(\sum X)(\sum Y)}{N}}{\sum X^2 - \frac{(\sum X)^2}{N}}$ | *Computational formula for slope of the linear regression line* |
| $\text{Chi-square} = \sum \frac{(O - E)^2}{E}$ | *Chi–square test* ($\chi^2$) |
| $\text{Chi-square} = \sum \frac{(\lvert O - E \rvert - .5)^2}{E}$ | *Chi–square test with correction for continuity* |
| $F_S = [F(df_{\text{Between}}, df_{\text{Within}})(a - 1)]$ | *Critical value of F for Scheffé's test* |
| $\binom{N}{f} = \frac{N!}{f!(N - f)!}$ | *Binomial coefficient* |

| | *Definition* |
|---|---|
| $p = \dfrac{(A+B)!(C+D)!(A+C)!(B+D)!}{N!A!B!C!D!}$ | *Fisher exact probability test* |
| $P(f) = \binom{N}{f} P^f Q^{N-f}$ | *Binomial test* |
| $r = \dfrac{SS_{XY}}{\sqrt{(SS_X)(SS_Y)}} = \dfrac{(\sum XY) - \dfrac{(\sum X)(\sum Y)}{N}}{\sqrt{\left(\sum X^2 - \dfrac{(\sum X)^2}{N}\right)\left(\sum Y^2 - \dfrac{(\sum Y)^2}{N}\right)}}$ | *Pearson correlation coefficient* |
| $\text{rho} = 1 - \dfrac{6(\sum D^2)}{n(n^2 - 1)}$ | *Spearman's rank difference correlation coefficient* |
| $s = \sqrt{s^2}$ | *Estimate of the standard deviation* |
| $s^2 = \dfrac{SS}{N-1}$ | *Variance estimate* |
| $s_{\bar{D}} = \dfrac{s_D}{\sqrt{N}}$ | *Standard error of the mean of difference scores* |
| $s_p^2 = \dfrac{SS_1 + SS_2}{N_1 + N_2 - 2}$ | *Pooled estimate of population variance based on two independent samples* |
| $s_{\bar{x}} = \dfrac{s}{\sqrt{N}}$ | *Standard error of the mean* |
| $s_{\bar{x}_1 - \bar{x}_2} = \sqrt{s_p^2\left(\dfrac{N_1 + N_2}{N_1 N_2}\right)}$ | *Standard error of the difference between two means* |
| $SS = \sum (X - \bar{X})^2$ | *Defining formula for sum of squares (sum of squared deviations from the mean)* |
| $SS = \sum X^2 - \dfrac{(\sum X)^2}{N}$ | *Computational formula for sum of squares (sum of squared deviations from the mean)* |
| $SS_A = \dfrac{\sum (A)^2}{bn} - \dfrac{(\sum X_T)^2}{N}$ | *Sum of squares for Factor A in a two-factor design (equal n's in each group)* |
| $SS_{AB} = \dfrac{\sum (AB)^2}{an} - \dfrac{\sum (A)^2}{bn} - \dfrac{\sum (B)^2}{an} + \dfrac{(\sum X_T)^2}{N}$ | *Sum of squares for the A × B interaction in a two-factor design (equal n's in each group)* |

| | Definition |
|---|---|
| $SS_B = \frac{\sum (B)^2}{an} - \frac{(\sum X_T)^2}{N}$ | *Sum of squares for Factor B in a two-factor design (equal n's in each group)* |
| $SS_{\text{Between}} = \frac{\sum (\sum X)^2}{n} - \frac{(\sum X_T)^2}{N}$ | *Sum of squares between groups in a one-factor design (equal n's in each group)* |
| $SS_{\text{Total}} = \sum X^2 - \frac{(\sum X_T)^2}{N}$ | *Total sum of squares in either a one- or two-factor design* |
| $SS_{\text{Within}} = \sum X^2 - \frac{\sum (AB)^2}{n}$ | *Sum of squares within groups in a two-factor design (equal n's in each group)* |
| $SS_{\text{Within}} = \sum X^2 - \frac{\sum (\sum X)^2}{n}$ | *Sum of squares within groups in a one-factor design (equal n's in each group)* |
| $t = \frac{\bar{D} - \mu_{\bar{D}}}{s_{\bar{D}}}$ | *t–test for non-independent groups (difference scores from within-subject or matched-pairs designs)* |
| $t = \frac{r - 0}{\sqrt{\frac{1 - r^2}{N - 2}}}$ | *t for testing the hypothesis that the population correlation coefficient equals zero* |
| $t = \frac{(\bar{X}_1 - \bar{X}_2) - (\mu_1 - \mu_2)}{s_{\bar{x}_1 - \bar{x}_2}}$ | *t–test for independent groups* |
| $U =$ the smaller of $U_1$ or $U_2$<br>$U_1 = n_1 n_2 + \frac{n_1(n_1 + 1)}{2} - \sum R_1$<br>$U_2 = n_1 n_2 + \frac{n_2(n_2 + 1)}{2} - \sum R_2$ | *Mann–Whitney U test* |
| $\bar{X} = \frac{\sum X}{N}$ | *Mean of a sample* |
| $Y' = a + bx$ | *Linear regression equation* |
| $z = \frac{\lvert f - NP \rvert - .5}{\sqrt{NPQ}}$ | *Normal approximation to the binomial* |
| $z = \frac{U - \frac{n_1 n_2}{2}}{\sqrt{\frac{(n_1)(n_2)(n_1 + n_2 + 1)}{12}}}$ | *Normal approximation for the Mann–Whitney U* |

# Appendixes

## APPENDIX A
## TABLE OF RANDOM NUMBERS

*Instructions for use:* The numbers in the table (on pages 212 and 213) have been generated by a computer so that every digit is as likely to appear as every other. If you want to select a sample of 100 cases from a population of 500, you would enter the table at any point (e.g., the upper right-hand corner) and then read three digit numbers until you had found 100 numbers between the values of 001 and 500. Any numbers you encountered which were larger than 500 would be ignored.

| | | | | | | | | | |
|---|---|---|---|---|---|---|---|---|---|
| 03991 | 10461 | 93716 | 16894 | 66083 | 24653 | 84609 | 58232 | 88618 | 19161 |
| 38555 | 95554 | 32886 | 59780 | 08355 | 60860 | 29735 | 47762 | 71299 | 23853 |
| 17546 | 73704 | 92052 | 46215 | 55121 | 29281 | 59076 | 07936 | 27954 | 58909 |
| 32643 | 52861 | 95819 | 06831 | 00911 | 98936 | 76355 | 93779 | 80863 | 00514 |
| 69572 | 68777 | 39510 | 35905 | 14060 | 40619 | 29549 | 69616 | 33564 | 60780 |
| 24122 | 66591 | 27699 | 06494 | 14845 | 46672 | 61958 | 77100 | 90899 | 75754 |
| 61196 | 30231 | 92962 | 61773 | 41839 | 55382 | 17267 | 70943 | 78038 | 70267 |
| 30532 | 21704 | 10274 | 12202 | 39685 | 23309 | 10061 | 68829 | 55986 | 66485 |
| 03788 | 97599 | 75867 | 20717 | 74416 | 53166 | 35208 | 33374 | 87539 | 08823 |
| 48228 | 63379 | 85783 | 47619 | 53152 | 67433 | 35663 | 52972 | 16818 | 60311 |
| 60365 | 94653 | 35075 | 33949 | 42614 | 29297 | 01918 | 28316 | 98953 | 73231 |
| 83799 | 42402 | 56623 | 34442 | 34994 | 41374 | 70071 | 14736 | 09958 | 18065 |
| 32960 | 07405 | 36409 | 83232 | 99385 | 41600 | 11133 | 07586 | 15917 | 06253 |
| 19322 | 53845 | 57620 | 52606 | 66497 | 68646 | 78138 | 66559 | 19640 | 99413 |
| 11220 | 94747 | 07399 | 37408 | 48509 | 23929 | 27482 | 45476 | 85244 | 35159 |
| 31751 | 57260 | 68980 | 05339 | 15470 | 48355 | 88651 | 22596 | 03152 | 19121 |
| 88492 | 99382 | 14454 | 04504 | 20094 | 98977 | 74843 | 93413 | 22109 | 78508 |
| 30934 | 47744 | 07481 | 83828 | 73788 | 06533 | 28597 | 20405 | 94205 | 20380 |
| 22888 | 48893 | 27499 | 98748 | 60530 | 45128 | 74022 | 84617 | 82037 | 10268 |
| 78212 | 16993 | 35902 | 91386 | 44372 | 15486 | 65741 | 14014 | 87481 | 37220 |
| 41849 | 84547 | 46850 | 52326 | 34677 | 58300 | 74910 | 64345 | 19325 | 81549 |
| 46352 | 33049 | 69248 | 93460 | 45305 | 07521 | 61318 | 31855 | 14413 | 70951 |
| 11087 | 96294 | 14013 | 31792 | 59747 | 67277 | 76503 | 34513 | 39663 | 77544 |
| 52701 | 08337 | 56303 | 87315 | 16520 | 69676 | 11654 | 99893 | 02181 | 68161 |
| 57275 | 36898 | 81304 | 48585 | 68652 | 27376 | 92852 | 55866 | 88448 | 03584 |
| 20857 | 73156 | 70284 | 24326 | 79375 | 95220 | 01159 | 63267 | 10622 | 48391 |
| 15633 | 84924 | 90415 | 93614 | 33521 | 26665 | 55823 | 47641 | 86225 | 31704 |
| 92694 | 48297 | 39904 | 02115 | 59589 | 49067 | 66821 | 41575 | 49767 | 04037 |
| 77613 | 19019 | 88152 | 00080 | 20554 | 91409 | 96277 | 48257 | 50816 | 97616 |
| 38688 | 32486 | 45134 | 63545 | 59404 | 72059 | 43947 | 51680 | 43852 | 59693 |
| 25163 | 01889 | 70014 | 15021 | 41290 | 67312 | 71857 | 15957 | 68971 | 11403 |
| 65251 | 07629 | 37239 | 33295 | 05870 | 01119 | 92784 | 26340 | 18477 | 65622 |
| 36815 | 43625 | 18637 | 37509 | 82444 | 99005 | 04921 | 73701 | 14707 | 93997 |
| 64397 | 11692 | 05327 | 82162 | 20247 | 81759 | 45197 | 25332 | 83745 | 22567 |
| 04515 | 25624 | 95096 | 67946 | 48460 | 85558 | 15191 | 18782 | 16930 | 33361 |
| 83761 | 60873 | 43253 | 84145 | 60833 | 25983 | 01291 | 41349 | 20368 | 07126 |
| 14387 | 06345 | 80854 | 09279 | 43529 | 06318 | 38384 | 74761 | 41196 | 37480 |
| 51321 | 92246 | 80088 | 77074 | 88722 | 56736 | 66164 | 49431 | 66919 | 31678 |
| 72472 | 00008 | 80890 | 18002 | 94813 | 31900 | 54155 | 83436 | 35352 | 54131 |
| 05466 | 55306 | 93128 | 18464 | 74457 | 90561 | 72848 | 11834 | 79982 | 68416 |
| 39528 | 72484 | 82474 | 25593 | 48545 | 35247 | 18619 | 13674 | 18611 | 19241 |
| 81616 | 18711 | 53342 | 44276 | 75122 | 11724 | 74627 | 73707 | 58319 | 15997 |
| 07586 | 16120 | 82641 | 22820 | 92904 | 13141 | 32392 | 19763 | 61199 | 67940 |
| 90767 | 04235 | 13574 | 17200 | 69902 | 63742 | 78464 | 22501 | 18627 | 90872 |
| 40188 | 28193 | 29593 | 88627 | 94972 | 11598 | 62095 | 36787 | 00441 | 58997 |
| 34414 | 82157 | 86887 | 55087 | 19152 | 00023 | 12302 | 80783 | 32624 | 68691 |
| 63439 | 75363 | 44989 | 16822 | 36024 | 00867 | 76378 | 41605 | 65961 | 73488 |
| 67049 | 09070 | 93399 | 45547 | 94458 | 74284 | 05041 | 49807 | 20288 | 34060 |
| 79495 | 04146 | 52162 | 90286 | 54158 | 34243 | 46978 | 35482 | 59362 | 95938 |
| 91704 | 30552 | 04737 | 21031 | 75051 | 93029 | 47665 | 64382 | 99782 | 93478 |

This table is reproduced with permission from the RAND Corporation: *A Million Random Digits*, 1955.

| | | | | | | | | | |
|---|---|---|---|---|---|---|---|---|---|
| 19612 | 78430 | 11661 | 94770 | 77603 | 65669 | 86868 | 12665 | 30012 | 75989 |
| 39141 | 77400 | 28000 | 64238 | 73258 | 71794 | 31340 | 26256 | 66453 | 37016 |
| 64756 | 80457 | 08747 | 12836 | 03469 | 50678 | 03274 | 43423 | 66677 | 82556 |
| 92901 | 51878 | 56441 | 22998 | 29718 | 38447 | 06453 | 25311 | 07565 | 53771 |
| 03551 | 90070 | 09483 | 94050 | 45938 | 18135 | 36908 | 43321 | 11073 | 51803 |
| 98884 | 66209 | 06830 | 53656 | 14663 | 56346 | 71430 | 04909 | 19818 | 05707 |
| 27369 | 86882 | 53473 | 07541 | 53633 | 70863 | 03748 | 12822 | 19360 | 49088 |
| 59066 | 75974 | 63335 | 20483 | 43514 | 37481 | 58278 | 26967 | 49325 | 43951 |
| 91647 | 93783 | 64169 | 49022 | 98588 | 09495 | 49829 | 59068 | 38831 | 04838 |
| 83605 | 92419 | 39542 | 07772 | 71568 | 75673 | 35185 | 89759 | 44901 | 74291 |
| 24895 | 88530 | 70774 | 35439 | 46758 | 70472 | 70207 | 92675 | 91623 | 61275 |
| 35720 | 26556 | 95596 | 20094 | 73750 | 85788 | 34264 | 01703 | 46833 | 65248 |
| 14141 | 53410 | 38649 | 06343 | 57256 | 61342 | 72709 | 75318 | 90379 | 37562 |
| 27416 | 75670 | 92176 | 72535 | 93119 | 56077 | 06886 | 18244 | 92344 | 31374 |
| 82071 | 07429 | 81007 | 47749 | 40744 | 56974 | 23336 | 88821 | 53841 | 10536 |
| 21445 | 82793 | 24831 | 93241 | 14199 | 76268 | 70883 | 68002 | 03829 | 17443 |
| 72513 | 76400 | 52225 | 92348 | 62308 | 98481 | 29744 | 33165 | 33141 | 61020 |
| 71479 | 45027 | 76160 | 57411 | 13780 | 13632 | 52308 | 77762 | 88874 | 33697 |
| 83210 | 51466 | 09088 | 50395 | 26743 | 05306 | 21706 | 70001 | 99439 | 80767 |
| 68749 | 95148 | 94897 | 78636 | 96750 | 09024 | 94538 | 91143 | 96693 | 61886 |
| 05184 | 75763 | 47075 | 88158 | 05313 | 53439 | 14908 | 08830 | 60096 | 21551 |
| 13651 | 62546 | 96892 | 25240 | 47511 | 58483 | 87342 | 78818 | 07855 | 39269 |
| 00566 | 21220 | 00292 | 24069 | 25072 | 29519 | 52548 | 54091 | 21282 | 21296 |
| 50958 | 17695 | 58072 | 68990 | 60329 | 95955 | 71586 | 63417 | 35947 | 67807 |
| 57621 | 64547 | 46850 | 37981 | 38527 | 09037 | 64756 | 03324 | 04986 | 83666 |
| 09282 | 25844 | 79139 | 78435 | 35428 | 43561 | 69799 | 63314 | 12991 | 93516 |
| 23394 | 94206 | 93432 | 37836 | 94919 | 26846 | 02555 | 74410 | 94915 | 48199 |
| 05280 | 37470 | 93622 | 04345 | 15092 | 19510 | 18094 | 16613 | 78234 | 50001 |
| 95491 | 97976 | 38306 | 32192 | 82639 | 54624 | 72434 | 92606 | 23191 | 74693 |
| 78521 | 00104 | 18248 | 75583 | 90326 | 50785 | 54034 | 66251 | 35774 | 14692 |
| 96345 | 44579 | 85932 | 44053 | 75704 | 20840 | 86583 | 83944 | 52456 | 73766 |
| 77963 | 31151 | 32364 | 91691 | 47357 | 40338 | 23435 | 24065 | 08458 | 95366 |
| 07520 | 11294 | 23238 | 01748 | 41690 | 67328 | 54814 | 37777 | 10057 | 42332 |
| 38423 | 02309 | 70703 | 85736 | 46148 | 14258 | 29236 | 12152 | 05088 | 65825 |
| 02463 | 65533 | 21199 | 60555 | 33928 | 01817 | 07396 | 89215 | 30722 | 22102 |
| 15880 | 92261 | 17292 | 88190 | 61781 | 48898 | 92525 | 21283 | 88581 | 60098 |
| 71926 | 00819 | 59144 | 00224 | 30570 | 90194 | 18329 | 06999 | 26857 | 19238 |
| 64425 | 28108 | 16554 | 16016 | 00042 | 83229 | 10333 | 36168 | 65617 | 94834 |
| 79782 | 23924 | 49440 | 30432 | 81077 | 31543 | 95216 | 64865 | 13658 | 51081 |
| 35337 | 74538 | 44553 | 64672 | 90960 | 41849 | 93865 | 44608 | 93176 | 34851 |
| 05249 | 29329 | 19715 | 94082 | 14738 | 86667 | 43708 | 66354 | 93692 | 25527 |
| 56463 | 99380 | 38793 | 85774 | 19056 | 13939 | 46062 | 27647 | 66146 | 63210 |
| 96296 | 33121 | 54196 | 34108 | 75814 | 85986 | 71171 | 15102 | 28992 | 63165 |
| 98380 | 36269 | 60014 | 07201 | 62448 | 46385 | 42175 | 88350 | 46182 | 49126 |
| 52567 | 64350 | 16315 | 53969 | 80395 | 81114 | 54358 | 64578 | 47269 | 15747 |
| 78498 | 90830 | 25955 | 99236 | 43286 | 91064 | 99969 | 95144 | 64424 | 77377 |
| 49553 | 24241 | 08150 | 89535 | 08703 | 91041 | 77323 | 81079 | 45127 | 93686 |
| 32151 | 07075 | 83155 | 10252 | 73100 | 88618 | 23891 | 87418 | 45417 | 20268 |
| 11314 | 50363 | 26860 | 27799 | 49416 | 83534 | 19187 | 08059 | 76677 | 02110 |
| 12364 | 71210 | 87052 | 50241 | 90785 | 97889 | 81399 | 58130 | 64439 | 05614 |

## APPENDIX B

Proof that the defining formula for the sum of squares equals the computational formula $\left[\sum (X - \bar{X})^2 = \sum X^2 - \frac{(\sum X)^2}{N}\right]$.

A. Mathematical Proof

$$\sum (X - \bar{X})^2 = \sum [X^2 - 2X\bar{X} + \bar{X}^2]$$

$$= \sum X^2 - \sum (2X\bar{X}) + \sum \bar{X}^2$$

$$= \sum X^2 - 2\bar{X}(\sum X) + \sum \bar{X}^2 \quad \text{(see point 1 below)}$$

$$= \sum X^2 - 2\bar{X}(\sum X) + N\bar{X}^2 \quad \text{(see point 2 below)}$$

$$= \sum X^2 - 2\left(\frac{\sum X}{N}\right)(\sum X) + N\left(\frac{\sum X}{N}\right)^2$$

(see point 3 below)

$$= \sum X^2 - 2\frac{(\sum X)^2}{N} + N\frac{(\sum X)^2}{N^2}$$

$$= \sum X^2 - 2\frac{(\sum X)^2}{N} + \frac{(\sum X)^2}{N}$$

$$\sum (X - \bar{X})^2 = \sum X^2 - \frac{(\sum X)^2}{N}$$

B. Example and Explanation

| $S$ | $X$ | $\bar{X}$ | $\bar{X}^2$ |
|---|---|---|---|
| 1 | 1 | 3 | 9 |
| 2 | 5 | 3 | 9 |
| 3 | 3 | 3 | 9 |

$$\sum X = 9$$

$$\bar{X} = 3.00$$

Remember, $\sum X = X_{(\text{for subject 1})} + X_{(\text{for subject 2})} + X_{(\text{for subject 3})}$

1) 2 and $\bar{X}$ are constants (that is, they have the same value for all subjects in the group) and, $\sum (\text{constant})X = \text{constant} \sum X$. For example,

$$\sum (2\bar{X}X) = (2\bar{X})(1) + (2\bar{X})(5) + (2\bar{X})(3) = 6(1) + 6(5) + 6(3) = 54$$

$$= 2\bar{X}(1 + 5 + 3) \qquad = 6(9) \qquad = 54$$

$$= 2\bar{X} \sum X$$

2) $\bar{X}^2$ is a constant (since it has the same value for all subjects) and

$$\sum (\text{constant}) = N(\text{constant}).$$

For example,

$$\begin{aligned}\sum \bar{X}^2 &= \bar{X}^2 + \bar{X}^2 + \bar{X}^2 = 9 + 9 + 9 = 27\\ &= 3\bar{X}^2 \qquad\qquad\quad = 3(9) \qquad\quad = 27\\ &= N\bar{X}^2\end{aligned}$$

3) Since $\bar{X} = \dfrac{\sum X}{N}$, $\dfrac{\sum X}{N}$ is substituted.

## APPENDIX C

## AREAS UNDER THE NORMAL CURVE (FOR $z$-SCORES)

*Instructions for use:* Find your obtained value of $z$ by first looking down the left-hand column until you find the value that corresponds to your obtained $z$ through the first decimal. Then move across this row until you enter the column that corresponds to the second digit after the decimal. You have now located the area between a $z$-score of 0 and your obtained score. For example, the proportion of the area under the curve corresponding to a $z$-score of $+1.96$ is found by locating 1.9 under column "$z$" and then moving across the table to the column headed ".—6." The value you have now located in the table should be .4750. Since the total area to the right of 0 is equal to .5000, you can obtain the area above $z = +1.96$ by subtracting .4750 from .5000. Thus, .0250 of the total area under the normal curve is beyond a $z$-score of $+1.96$. Similarly, .4750 of the area is between a $z$-score of 0 and $-1.96$, inasmuch as the curve is symmetrical.

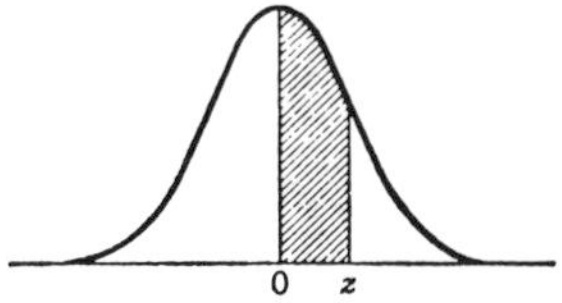


| z | .–0 | .–1 | .–2 | .–3 | .–4 | .–5 | .–6 | .–7 | .–8 | .–9 |
|---|---|---|---|---|---|---|---|---|---|---|
| 0.0 | .0000 | .0040 | .0080 | .0120 | .0160 | .0199 | .0239 | .0279 | .0319 | .0359 |
| 0.1 | .0398 | .0438 | .0478 | .0517 | .0557 | .0596 | .0636 | .0675 | .0714 | .0753 |
| 0.2 | .0793 | .0832 | .0871 | .0910 | .0948 | .0987 | .1026 | .1064 | .1103 | .1141 |
| 0.3 | .1179 | .1217 | .1255 | .1293 | .1331 | .1368 | .1406 | .1443 | .1480 | .1517 |
| 0.4 | .1554 | .1591 | .1628 | .1664 | .1700 | .1736 | .1772 | .1808 | .1844 | .1879 |
| 0.5 | .1915 | .1950 | .1985 | .2019 | .2054 | .2088 | .2123 | .2157 | .2190 | .2224 |
| 0.6 | .2257 | .2291 | .2324 | .2357 | .2389 | .2422 | .2454 | .2486 | .2517 | .2549 |
| 0.7 | .2580 | .2611 | .2642 | .2673 | .2703 | .2734 | .2764 | .2794 | .2823 | .2852 |
| 0.8 | .2881 | .2910 | .2939 | .2967 | .2995 | .3023 | .3051 | .3078 | .3106 | .3133 |
| 0.9 | .3159 | .3186 | .3212 | .3238 | .3264 | .3289 | .3315 | .3340 | .3365 | .3389 |
| 1.0 | .3413 | .3438 | .3461 | .3485 | .3508 | .3531 | .3554 | .3577 | .3599 | .3621 |
| 1.1 | .3643 | .3665 | .3686 | .3708 | .3729 | .3749 | .3770 | .3790 | .3810 | .3830 |
| 1.2 | .3849 | .3869 | .3888 | .3907 | .3925 | .3944 | .3962 | .3980 | .3997 | .4015 |
| 1.3 | .4032 | .4049 | .4066 | .4082 | .4099 | .4115 | .4131 | .4147 | .4162 | .4177 |
| 1.4 | .4192 | .4207 | .4222 | .4236 | .4251 | .4265 | .4279 | .4292 | .4306 | .4319 |
| 1.5 | .4332 | .4345 | .4357 | .4370 | .4382 | .4394 | .4406 | .4418 | .4429 | .4441 |
| 1.6 | .4452 | .4463 | .4474 | .4484 | .4495 | .4505 | .4515 | .4525 | .4535 | .4545 |
| 1.7 | .4554 | .4564 | .4573 | .4582 | .4591 | .4599 | .4608 | .4616 | .4625 | .4633 |
| 1.8 | .4641 | .4649 | .4656 | .4664 | .4671 | .4678 | .4686 | .4693 | .4699 | .4706 |
| 1.9 | .4713 | .4719 | .4726 | .4732 | .4738 | .4744 | .4750 | .4756 | .4761 | .4767 |
| 2.0 | .4772 | .4778 | .4783 | .4788 | .4793 | .4798 | .4803 | .4808 | .4812 | .4817 |
| 2.1 | .4821 | .4826 | .4830 | .4834 | .4838 | .4842 | .4846 | 4850 | .4854 | .4857 |
| 2.2 | .4861 | .4864 | .4868 | .4871 | .4875 | .4878 | .4881 | .4884 | .4887 | .4890 |
| 2.3 | .4893 | .4896 | .4898 | .4901 | .4904 | .4906 | .4909 | .4911 | .4913 | .4916 |
| 2.4 | .4918 | .4920 | .4922 | .4925 | .4927 | .4929 | .4931 | .4932 | .4934 | .4936 |
| 2.5 | .4938 | .4940 | .4941 | .4943 | .4945 | .4946 | .4948 | .4949 | .4951 | .4952 |
| 2.6 | .4953 | .4955 | .4956 | .4957 | .4959 | .4960 | .4961 | .4962 | .4963 | .4964 |
| 2.7 | .4965 | .4966 | .4967 | .4968 | .4969 | .4970 | .4971 | .4972 | .4973 | .4974 |
| 2.8 | .4974 | .4975 | .4976 | .4977 | .4977 | .4978 | .4979 | .4979 | .4980 | .4981 |
| 2.9 | .4981 | .4982 | .4982 | .4983 | .4984 | .4984 | .4985 | .4985 | .4986 | .4986 |
| 3.0 | .4987 | .4987 | .4987 | .4988 | .4988 | .4989 | .4989 | .4989 | .4990 | .4990 |

Reproduced by permission from P. G. Hoel, *Elementary Statistics*, 2nd Ed. New York: John Wiley and Sons, Inc., 1966.

## APPENDIX D
## CRITICAL VALUES OF $t$

*Instructions for use:* Locate the degrees of freedom ($df$) appropriate to your study in the left-hand column. Then move across this row until you enter the column that corresponds to the proportion of the curve beyond which you would reject $H_0$. For example, with $df = 7$ a value of $t$ above $+2.365$ would occur approximately 2.5 times in 100 (.025). A value below $-2.365$ would also occur approximately 2.5 times in 100, because the curve is symmetrical. Therefore, with an alpha level equal to .05, your critical value of $t$ would be 2.365. Values above $+2.365$ and values below $-2.365$ would lead you to reject $H_0$ (when using a two-tailed test). With $df = 7$ and alpha $= .05$, for a one-tailed test, the critical value of $t = 1.895$.

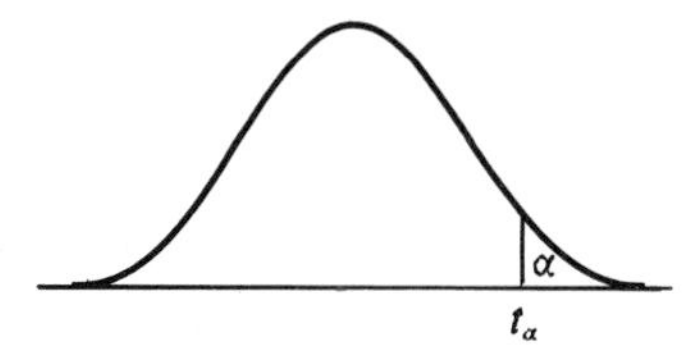


| $df$ | alpha levels for one-tailed tests $t_{.100}$ | $t_{.050}$ | $t_{.025}$ | $t_{.010}$ | $t_{.005}$ |
|---|---|---|---|---|---|
| **1** | 3.078 | 6.314 | 12.706 | 31.821 | 63.657 |
| **2** | 1.886 | 2.920 | 4.303 | 6.965 | 9.925 |
| **3** | 1.638 | 2.353 | 3.182 | 4.541 | 5.841 |
| **4** | 1.533 | 2.132 | 2.776 | 3.747 | 4.604 |
| **5** | 1.476 | 2.015 | 2.571 | 3.365 | 4.032 |
| **6** | 1.440 | 1.943 | 2.447 | 3.143 | 3.707 |
| **7** | 1.415 | 1.895 | 2.365 | 2.998 | 3.499 |
| **8** | 1.397 | 1.860 | 2.306 | 2.896 | 3.355 |
| **9** | 1.383 | 1.833 | 2.262 | 2.821 | 3.250 |
| **10** | 1.372 | 1.812 | 2.228 | 2.764 | 3.169 |
| **11** | 1.363 | 1.796 | 2.201 | 2.718 | 3.106 |
| **12** | 1.356 | 1.782 | 2.179 | 2.681 | 3.055 |
| **13** | 1.350 | 1.771 | 2.160 | 2.650 | 3.012 |
| **14** | 1.345 | 1.761 | 2.145 | 2.624 | 2.977 |
| **15** | 1.341 | 1.753 | 2.131 | 2.602 | 2.947 |
| **16** | 1.337 | 1.746 | 2.120 | 2.583 | 2.921 |
| **17** | 1.333 | 1.740 | 2.110 | 2.567 | 2.898 |
| **18** | 1.330 | 1.734 | 2.101 | 2.552 | 2.878 |
| **19** | 1.328 | 1.729 | 2.093 | 2.539 | 2.861 |
| **20** | 1.325 | 1.725 | 2.086 | 2.528 | 2.845 |
| **21** | 1.323 | 1.721 | 2.080 | 2.518 | 2.831 |
| **22** | 1.321 | 1.717 | 2.074 | 2.508 | 2.819 |
| **23** | 1.319 | 1.714 | 2.069 | 2.500 | 2.807 |
| **24** | 1.318 | 1.711 | 2.064 | 2.492 | 2.797 |
| **25** | 1.316 | 1.708 | 2.060 | 2.485 | 2.787 |
| **26** | 1.315 | 1.706 | 2.056 | 2.479 | 2.779 |
| **27** | 1.314 | 1.703 | 2.052 | 2.473 | 2.771 |
| **28** | 1.313 | 1.701 | 2.048 | 2.467 | 2.763 |
| **29** | 1.311 | 1.699 | 2.045 | 2.462 | 2.756 |
| **inf.** | 1.282 | 1.645 | 1.960 | 2.326 | 2.576 |
| | $t_{.200}$ | $t_{.100}$ | $t_{.050}$ | $t_{.020}$ | $t_{.010}$ |
| | alpha levels for two-tailed tests | | | | |

Adapted from tables computed by Maxine Merrington in *Biometrika*, Vol. 32, 1941, p. 300, for Prof. E. S. Pearson, and Oliver and Boyd, *Biometrika Tables for Statisticians,* Vol. 1, 3rd. ed., 1966, by permission of the authors.

# APPENDIX E
## CRITICAL VALUES OF $F$

*Instructions for use:* Find the column containing the number of degrees of freedom for the numerator and the row containing the number of degrees of freedom for the denominator. If the value in the body of the table in regular type is exceeded by the obtained $F$ value, the 5 percent significance level has been reached. For example, an $F$ of 5.02 with *df* for the numerator equal to 1 and *df* for the denominator equal to 10 exceeds the table value of 4.96 at the 5 percent level and would be judged significant at that level. The value in boldface type is the critical value of $F$ at the .01 level. Thus, with 1 and 10 *df*, the obtained value would have to be greater than 10.04 to be considered significant.

Degrees of freedom for numerator (columns); Degrees of freedom for denominator (rows)

| Degrees of freedom for denominator | 1 | 2 | 3 | 4 | 5 | 6 | 7 | 8 | 9 | 10 | 11 | 12 |
|---|---|---|---|---|---|---|---|---|---|---|---|---|
| 1 | 161<br>**4,052** | 200<br>**4,999** | 216<br>**5,403** | 225<br>**5,625** | 230<br>**5,764** | 234<br>**5,859** | 237<br>**5,928** | 239<br>**5,981** | 241<br>**6,022** | 242<br>**6,056** | 243<br>**6,082** | 244<br>**6,106** |
| 2 | 18.51<br>**98.49** | 19.00<br>**99.00** | 19.16<br>**99.17** | 19.25<br>**99.25** | 19.30<br>**99.30** | 19.33<br>**99.33** | 19.36<br>**99.36** | 19.37<br>**99.37** | 19.38<br>**99.39** | 19.39<br>**99.40** | 19.40<br>**99.41** | 19.41<br>**99.42** |
| 3 | 10.13<br>**34.12** | 9.55<br>**30.82** | 9.28<br>**29.46** | 9.12<br>**28.71** | 9.01<br>**28.24** | 8.94<br>**27.91** | 8.88<br>**27.67** | 8.84<br>**27.49** | 8.81<br>**27.34** | 8.78<br>**27.23** | 8.76<br>**27.13** | 8.74<br>**27.05** |
| 4 | 7.71<br>**21.20** | 6.94<br>**18.00** | 6.59<br>**16.69** | 6.39<br>**15.98** | 6.26<br>**15.52** | 6.16<br>**15.21** | 6.09<br>**14.98** | 6.04<br>**14.80** | 6.00<br>**14.66** | 5.96<br>**14.54** | 5.93<br>**14.45** | 5.91<br>**14.37** |
| 5 | 6.61<br>**16.26** | 5.79<br>**13.27** | 5.41<br>**12.06** | 5.19<br>**11.39** | 5.05<br>**10.97** | 4.95<br>**10.67** | 4.88<br>**10.45** | 4.82<br>**10.29** | 4.78<br>**10.15** | 4.74<br>**10.05** | 4.70<br>**9.96** | 4.68<br>**9.89** |
| 6 | 5.99<br>**13.74** | 5.14<br>**10.92** | 4.76<br>**9.78** | 4.53<br>**9.15** | 4.39<br>**8.75** | 4.28<br>**8.47** | 4.21<br>**8.26** | 4.15<br>**8.10** | 4.10<br>**7.98** | 4.06<br>**7.87** | 4.03<br>**7.79** | 4.00<br>**7.72** |
| 7 | 5.59<br>**12.25** | 4.74<br>**9.55** | 4.35<br>**8.45** | 4.12<br>**7.85** | 3.97<br>**7.46** | 3.87<br>**7.19** | 3.79<br>**7.00** | 3.73<br>**6.84** | 3.68<br>**6.71** | 3.63<br>**6.62** | 3.60<br>**6.54** | 3.57<br>**6.47** |
| 8 | 5.32<br>**11.26** | 4.46<br>**8.65** | 4.07<br>**7.59** | 3.84<br>**7.01** | 3.69<br>**6.63** | 3.58<br>**6.37** | 3.50<br>**6.19** | 3.44<br>**6.03** | 3.39<br>**5.91** | 3.34<br>**5.82** | 3.31<br>**5.74** | 3.28<br>**5.67** |
| 9 | 5.12<br>**10.56** | 4.26<br>**8.02** | 3.86<br>**6.99** | 3.63<br>**6.42** | 3.48<br>**6.06** | 3.37<br>**5.80** | 3.29<br>**5.62** | 3.23<br>**5.47** | 3.18<br>**5.35** | 3.13<br>**5.26** | 3.10<br>**5.18** | 3.07<br>**5.11** |
| 10 | 4.96<br>**10.04** | 4.10<br>**7.56** | 3.71<br>**6.55** | 3.48<br>**5.99** | 3.33<br>**5.64** | 3.22<br>**5.39** | 3.14<br>**5.21** | 3.07<br>**5.06** | 3.02<br>**4.95** | 2.97<br>**4.85** | 2.94<br>**4.78** | 2.91<br>**4.71** |

| Degrees of freedom for denominator | 14 | 16 | 20 | 24 | 30 | 40 | 50 | 75 | 100 | 200 | 500 | ∞ |
|---|---|---|---|---|---|---|---|---|---|---|---|---|
| 1 | 245<br>**6,142** | 246<br>**6,169** | 248<br>**6,208** | 249<br>**6,234** | 250<br>**6,261** | 251<br>**6,286** | 252<br>**6,302** | 253<br>**6,323** | 253<br>**6,334** | 254<br>**6,352** | 254<br>**6,361** | 254<br>**6,366** |
| 2 | 19.42<br>**99.43** | 19.43<br>**99.44** | 19.44<br>**99.45** | 19.45<br>**99.46** | 19.46<br>**99.47** | 19.47<br>**99.48** | 19.47<br>**99.48** | 19.48<br>**99.49** | 19.49<br>**99.49** | 19.49<br>**99.49** | 19.50<br>**99.50** | 19.50<br>**99.50** |
| 3 | 8.71<br>**26.92** | 8.69<br>**26.83** | 8.66<br>**26.69** | 8.64<br>**26.60** | 8.62<br>**26.50** | 8.60<br>**26.41** | 8.58<br>**26.35** | 8.57<br>**26.27** | 8.56<br>**26.23** | 8.54<br>**26.18** | 8.54<br>**26.14** | 8.53<br>**26.12** |
| 4 | 5.87<br>**14.24** | 5.84<br>**14.15** | 5.80<br>**14.02** | 5.77<br>**13.93** | 5.74<br>**13.83** | 5.71<br>**13.74** | 5.70<br>**13.69** | 5.68<br>**13.61** | 5.66<br>**13.57** | 5.65<br>**13.52** | 5.64<br>**13.48** | 5.63<br>**13.46** |
| 5 | 4.64<br>**9.77** | 4.60<br>**9.68** | 4.56<br>**9.55** | 4.53<br>**9.47** | 4.50<br>**9.38** | 4.46<br>**9.29** | 4.44<br>**9.24** | 4.42<br>**9.17** | 4.40<br>**9.13** | 4.38<br>**9.07** | 4.37<br>**9.04** | 4.36<br>**9.02** |
| 6 | 3.96<br>**7.60** | 3.92<br>**7.52** | 3.87<br>**7.39** | 3.84<br>**7.31** | 3.81<br>**7.23** | 3.77<br>**7.14** | 3.75<br>**7.09** | 3.72<br>**7.02** | 3.71<br>**6.99** | 3.69<br>**6.94** | 3.68<br>**6.90** | 3.67<br>**6.88** |
| 7 | 3.52<br>**6.35** | 3.49<br>**6.27** | 3.44<br>**6.15** | 3.41<br>**6.07** | 3.38<br>**5.98** | 3.34<br>**5.90** | 3.32<br>**5.85** | 3.29<br>**5.78** | 3.28<br>**5.75** | 3.25<br>**5.70** | 3.24<br>**5.67** | 3.23<br>**5.65** |
| 8 | 3.23<br>**5.56** | 3.20<br>**5.48** | 3.15<br>**5.36** | 3.12<br>**5.28** | 3.08<br>**5.20** | 3.05<br>**5.11** | 3.03<br>**5.06** | 3.00<br>**5.00** | 2.98<br>**4.96** | 2.96<br>**4.91** | 2.94<br>**4.88** | 2.93<br>**4.86** |
| 9 | 3.02<br>**5.00** | 2.98<br>**4.92** | 2.93<br>**4.80** | 2.90<br>**4.73** | 2.86<br>**4.64** | 2.82<br>**4.56** | 2.80<br>**4.51** | 2.77<br>**4.45** | 2.76<br>**4.41** | 2.73<br>**4.36** | 2.72<br>**4.33** | 2.71<br>**4.31** |
| 10 | 2.86<br>**4.60** | 2.82<br>**4.52** | 2.77<br>**4.41** | 2.74<br>**4.33** | 2.70<br>**4.25** | 2.67<br>**4.17** | 2.64<br>**4.12** | 2.61<br>**4.05** | 2.59<br>**4.01** | 2.56<br>**3.96** | 2.55<br>**3.93** | 2.54<br>**3.91** |

| Degrees of freedom for denominator | Degrees of freedom for numerator | | | | | | | | | | | | | | | | | | | | | | | | Degrees of freedom for denominator |
|---|---|---|---|---|---|---|---|---|---|---|---|---|---|---|---|---|---|---|---|---|---|---|---|---|---|
| | 1 | 2 | 3 | 4 | 5 | 6 | 7 | 8 | 9 | 10 | 11 | 12 | 14 | 16 | 20 | 24 | 30 | 40 | 50 | 75 | 100 | 200 | 500 | ∞ | |
| 11 | 4.84<br>**9.65** | 3.98<br>**7.20** | 3.59<br>**6.22** | 3.36<br>**5.67** | 3.20<br>**5.32** | 3.09<br>**5.07** | 3.01<br>**4.88** | 2.95<br>**4.74** | 2.90<br>**4.63** | 2.86<br>**4.54** | 2.82<br>**4.46** | 2.79<br>**4.40** | 2.74<br>**4.29** | 2.70<br>**4.21** | 2.65<br>**4.10** | 2.61<br>**4.02** | 2.57<br>**3.94** | 2.53<br>**3.86** | 2.50<br>**3.80** | 2.47<br>**3.74** | 2.45<br>**3.70** | 2.42<br>**3.66** | 2.41<br>**3.62** | 2.40<br>**3.60** | 11 |
| 12 | 4.75<br>**9.33** | 3.88<br>**6.93** | 3.49<br>**5.95** | 3.26<br>**5.41** | 3.11<br>**5.06** | 3.00<br>**4.82** | 2.92<br>**4.65** | 2.85<br>**4.50** | 2.80<br>**4.39** | 2.76<br>**4.30** | 2.72<br>**4.22** | 2.69<br>**4.16** | 2.64<br>**4.05** | 2.60<br>**3.98** | 2.54<br>**3.86** | 2.50<br>**3.78** | 2.46<br>**3.70** | 2.42<br>**3.61** | 2.40<br>**3.56** | 2.36<br>**3.49** | 2.35<br>**3.46** | 2.32<br>**3.41** | 2.31<br>**3.38** | 2.30<br>**3.36** | 12 |
| 13 | 4.67<br>**9.07** | 3.80<br>**6.70** | 3.41<br>**5.74** | 3.18<br>**5.20** | 3.02<br>**4.86** | 2.92<br>**4.62** | 2.84<br>**4.44** | 2.77<br>**4.30** | 2.72<br>**4.19** | 2.67<br>**4.10** | 2.63<br>**4.02** | 2.60<br>**3.96** | 2.55<br>**3.85** | 2.51<br>**3.78** | 2.46<br>**3.67** | 2.42<br>**3.59** | 2.38<br>**3.51** | 2.34<br>**3.42** | 2.32<br>**3.37** | 2.28<br>**3.30** | 2.26<br>**3.27** | 2.24<br>**3.21** | 2.22<br>**3.18** | 2.21<br>**3.16** | 13 |
| 14 | 4.60<br>**8.86** | 3.74<br>**6.51** | 3.34<br>**5.56** | 3.11<br>**5.03** | 2.96<br>**4.69** | 2.85<br>**4.46** | 2.77<br>**4.28** | 2.70<br>**4.14** | 2.65<br>**4.03** | 2.60<br>**3.94** | 2.56<br>**3.86** | 2.53<br>**3.80** | 2.48<br>**3.70** | 2.44<br>**3.62** | 2.39<br>**3.51** | 2.35<br>**3.43** | 2.31<br>**3.34** | 2.27<br>**3.26** | 2.24<br>**3.21** | 2.21<br>**3.14** | 2.19<br>**3.11** | 2.16<br>**3.06** | 2.14<br>**3.02** | 2.13<br>**3.00** | 14 |
| 15 | 4.54<br>**8.68** | 3.68<br>**6.36** | 3.29<br>**5.42** | 3.06<br>**4.89** | 2.90<br>**4.56** | 2.79<br>**4.32** | 2.70<br>**4.14** | 2.64<br>**4.00** | 2.59<br>**3.89** | 2.55<br>**3.80** | 2.51<br>**3.73** | 2.48<br>**3.67** | 2.43<br>**3.56** | 2.39<br>**3.48** | 2.33<br>**3.36** | 2.29<br>**3.29** | 2.25<br>**3.20** | 2.21<br>**3.12** | 2.18<br>**3.07** | 2.15<br>**3.00** | 2.12<br>**2.97** | 2.10<br>**2.92** | 2.08<br>**2.89** | 2.07<br>**2.87** | 15 |
| 16 | 4.49<br>**8.53** | 3.63<br>**6.23** | 3.24<br>**5.29** | 3.01<br>**4.77** | 2.85<br>**4.44** | 2.74<br>**4.20** | 2.66<br>**4.03** | 2.59<br>**3.89** | 2.54<br>**3.78** | 2.49<br>**3.69** | 2.45<br>**3.61** | 2.42<br>**3.55** | 2.37<br>**3.45** | 2.33<br>**3.37** | 2.28<br>**3.25** | 2.24<br>**3.18** | 2.20<br>**3.10** | 2.16<br>**3.01** | 2.13<br>**2.96** | 2.09<br>**2.98** | 2.07<br>**2.86** | 2.04<br>**2.80** | 2.02<br>**2.77** | 2.01<br>**2.75** | 16 |
| 17 | 4.45<br>**8.40** | 3.59<br>**6.11** | 3.20<br>**5.18** | 2.96<br>**4.67** | 2.81<br>**4.34** | 2.70<br>**4.10** | 2.62<br>**3.93** | 2.55<br>**3.79** | 2.50<br>**3.68** | 2.45<br>**3.59** | 2.41<br>**3.52** | 2.38<br>**3.45** | 2.33<br>**3.35** | 2.29<br>**3.27** | 2.23<br>**3.16** | 2.19<br>**3.08** | 2.15<br>**3.00** | 2.11<br>**2.92** | 2.08<br>**2.86** | 2.04<br>**2.79** | 2.02<br>**2.76** | 1.99<br>**2.70** | 1.97<br>**2.67** | 1.96<br>**2.65** | 17 |
| 18 | 4.41<br>**8.28** | 3.55<br>**6.01** | 3.16<br>**5.09** | 2.93<br>**4.58** | 2.77<br>**4.25** | 3.66<br>**4.01** | 2.58<br>**3.85** | 2.51<br>**3.71** | 2.46<br>**3.60** | 2.41<br>**3.51** | 2.37<br>**3.44** | 2.34<br>**3.37** | 2.29<br>**3.27** | 2.25<br>**3.19** | 2.19<br>**3.07** | 2.15<br>**3.00** | 2.11<br>**2.91** | 2.07<br>**2.83** | 2.04<br>**2.78** | 2.00<br>**2.71** | 1.98<br>**2.68** | 1.95<br>**2.62** | 1.93<br>**2.59** | 1.92<br>**2.57** | 18 |
| 19 | 4.38<br>**8.18** | 3.52<br>**5.93** | 3.13<br>**5.01** | 2.90<br>**4.50** | 2.74<br>**4.17** | 2.63<br>**3.94** | 2.55<br>**3.77** | 2.48<br>**3.63** | 2.43<br>**3.52** | 2.38<br>**3.43** | 2.34<br>**3.36** | 2.31<br>**3.30** | 2.26<br>**3.19** | 2.21<br>**3.12** | 2.15<br>**3.00** | 2.11<br>**2.92** | 2.07<br>**2.84** | 2.02<br>**2.76** | 2.00<br>**2.70** | 1.96<br>**2.63** | 1.94<br>**2.60** | 1.91<br>**2.54** | 1.90<br>**2.51** | 1.88<br>**2.49** | 19 |
| 20 | 4.35<br>**8.10** | 3.49<br>**5.85** | 3.10<br>**4.94** | 2.87<br>**4.43** | 2.71<br>**4.10** | 2.60<br>**3.87** | 2.52<br>**3.71** | 2.45<br>**3.56** | 2.40<br>**3.45** | 2.35<br>**3.37** | 2.31<br>**3.30** | 2.28<br>**3.23** | 2.23<br>**3.13** | 2.18<br>**3.05** | 2.12<br>**2.94** | 2.08<br>**2.86** | 2.04<br>**2.77** | 1.99<br>**2.69** | 1.96<br>**2.63** | 1.92<br>**2.56** | 1.90<br>**2.53** | 1.87<br>**2.47** | 1.85<br>**2.44** | 1.84<br>**2.42** | 20 |
| 21 | 4.32<br>**8.02** | 3.47<br>**5.78** | 3.07<br>**4.87** | 2.84<br>**4.37** | 2.68<br>**4.04** | 2.57<br>**3.81** | 2.49<br>**3.65** | 2.42<br>**3.51** | 2.37<br>**3.40** | 2.32<br>**3.31** | 2.28<br>**3.24** | 2.25<br>**3.17** | 2.20<br>**3.07** | 2.15<br>**2.99** | 2.09<br>**2.88** | 2.05<br>**2.80** | 2.00<br>**2.72** | 1.96<br>**2.63** | 1.93<br>**2.58** | 1.89<br>**2.51** | 1.87<br>**2.47** | 1.84<br>**2.42** | 1.82<br>**2.38** | 1.81<br>**2.36** | 21 |
| 22 | 4.30<br>**7.94** | 3.44<br>**5.72** | 3.05<br>**4.82** | 2.82<br>**4.31** | 2.66<br>**3.99** | 2.55<br>**3.76** | 2.47<br>**3.59** | 2.40<br>**3.45** | 2.35<br>**3.35** | 2.30<br>**3.26** | 2.26<br>**3.18** | 2.23<br>**3.12** | 2.18<br>**3.02** | 2.13<br>**2.94** | 2.07<br>**2.83** | 2.03<br>**2.75** | 1.98<br>**2.67** | 1.93<br>**2.58** | 1.91<br>**2.53** | 1.87<br>**2.46** | 1.84<br>**2.42** | 1.81<br>**2.37** | 1.80<br>**2.33** | 1.78<br>**2.31** | 22 |
| 23 | 4.28<br>**7.88** | 3.42<br>**5.66** | 3.03<br>**4.76** | 2.80<br>**4.26** | 2.64<br>**3.94** | 2.53<br>**3.71** | 2.45<br>**3.54** | 2.38<br>**3.41** | 2.32<br>**3.30** | 2.28<br>**3.21** | 2.24<br>**3.14** | 2.20<br>**3.07** | 2.14<br>**2.97** | 2.10<br>**2.89** | 2.04<br>**2.78** | 2.00<br>**2.70** | 1.96<br>**2.62** | 1.91<br>**2.53** | 1.88<br>**2.48** | 1.84<br>**2.41** | 1.82<br>**2.37** | 1.79<br>**2.32** | 1.77<br>**2.28** | 1.76<br>**2.26** | 23 |
| 24 | 4.26<br>**7.82** | 3.40<br>**5.61** | 3.01<br>**4.72** | 2.78<br>**4.22** | 2.62<br>**3.90** | 2.51<br>**3.67** | 2.43<br>**3.50** | 2.36<br>**3.36** | 2.30<br>**3.25** | 2.26<br>**3.17** | 2.22<br>**3.09** | 2.18<br>**3.03** | 2.13<br>**2.93** | 2.09<br>**2.85** | 2.02<br>**2.74** | 1.98<br>**2.66** | 1.94<br>**2.58** | 1.89<br>**2.49** | 1.86<br>**2.44** | 1.82<br>**2.36** | 1.80<br>**2.33** | 1.76<br>**2.27** | 1.74<br>**2.23** | 1.73<br>**2.21** | 24 |
| 25 | 4.24<br>**7.77** | 3.38<br>**5.57** | 2.99<br>**4.68** | 2.76<br>**4.18** | 2.60<br>**3.86** | 2.49<br>**3.63** | 2.41<br>**3.46** | 2.34<br>**3.32** | 2.28<br>**3.21** | 2.24<br>**3.13** | 2.20<br>**3.05** | 2.16<br>**2.99** | 2.11<br>**2.89** | 2.06<br>**2.81** | 2.00<br>**2.70** | 1.96<br>**2.62** | 1.92<br>**2.54** | 1.87<br>**2.45** | 1.84<br>**2.40** | 1.80<br>**2.32** | 1.77<br>**2.29** | 1.74<br>**2.23** | 1.72<br>**2.19** | 1.71<br>**2.17** | 25 |

*Table continued on the following page*

| Degrees of freedom for denominator | Degrees of freedom for numerator | | | | | | | | | | | | | | | | | | | | | | | | Degrees of freedom for denominator |
|---|---|---|---|---|---|---|---|---|---|---|---|---|---|---|---|---|---|---|---|---|---|---|---|---|---|
| | 1 | 2 | 3 | 4 | 5 | 6 | 7 | 8 | 9 | 10 | 11 | 12 | 14 | 16 | 20 | 24 | 30 | 40 | 50 | 75 | 100 | 200 | 500 | $\infty$ | |
| 26 | 4.22<br>**7.72** | 3.37<br>**5.53** | 2.98<br>**4.64** | 2.74<br>**4.14** | 2.59<br>**3.82** | 2.47<br>**3.59** | 2.39<br>**3.42** | 2.32<br>**3.29** | 2.27<br>**3.17** | 2.22<br>**3.09** | 2.18<br>**3.02** | 2.15<br>**2.96** | 2.10<br>**2.86** | 2.05<br>**2.77** | 1.99<br>**2.66** | 1.95<br>**2.58** | 1.90<br>**2.50** | 1.85<br>**2.41** | 1.82<br>**2.36** | 1.78<br>**2.28** | 1.76<br>**2.25** | 1.72<br>**2.19** | 1.70<br>**2.15** | 1.69<br>**2.13** | 26 |
| 27 | 4.21<br>**7.68** | 3.35<br>**5.49** | 2.96<br>**4.60** | 2.73<br>**4.11** | 2.57<br>**3.79** | 2.46<br>**3.56** | 2.37<br>**3.39** | 2.30<br>**3.26** | 2.25<br>**3.14** | 2.20<br>**3.06** | 2.16<br>**2.98** | 2.13<br>**2.93** | 2.08<br>**2.83** | 2.03<br>**2.74** | 1.97<br>**2.63** | 1.93<br>**2.55** | 1.88<br>**2.47** | 1.84<br>**2.38** | 1.80<br>**2.33** | 1.76<br>**2.25** | 1.74<br>**2.21** | 1.71<br>**2.16** | 1.68<br>**2.12** | 1.67<br>**2.10** | 27 |
| 28 | 4.20<br>**7.64** | 3.34<br>**5.45** | 2.95<br>**4.57** | 2.71<br>**4.07** | 2.56<br>**3.76** | 2.44<br>**3.53** | 2.36<br>**3.36** | 2.29<br>**3.23** | 2.24<br>**3.11** | 2.19<br>**3.03** | 2.15<br>**2.95** | 2.12<br>**2.90** | 2.06<br>**2.80** | 2.02<br>**2.71** | 1.96<br>**2.60** | 1.91<br>**2.52** | 1.87<br>**2.44** | 1.81<br>**2.35** | 1.78<br>**2.30** | 1.75<br>**2.22** | 1.72<br>**2.18** | 1.69<br>**2.13** | 1.67<br>**2.09** | 1.65<br>**2.06** | 28 |
| 29 | 4.18<br>**7.60** | 3.33<br>**5.42** | 2.93<br>**4.54** | 2.70<br>**4.04** | 2.54<br>**3.73** | 2.43<br>**3.50** | 2.35<br>**3.33** | 2.28<br>**3.20** | 2.22<br>**3.08** | 2.18<br>**3.00** | 2.14<br>**2.92** | 2.10<br>**2.87** | 2.05<br>**2.77** | 2.00<br>**2.68** | 1.94<br>**2.57** | 1.90<br>**2.49** | 1.85<br>**2.41** | 1.80<br>**2.32** | 1.77<br>**2.27** | 1.73<br>**2.19** | 1.71<br>**2.15** | 1.68<br>**2.10** | 1.65<br>**2.06** | 1.64<br>**2.03** | 29 |
| 30 | 4.17<br>**7.56** | 3.32<br>**5.39** | 2.92<br>**4.51** | 2.69<br>**4.02** | 2.53<br>**3.70** | 2.42<br>**3.47** | 2.34<br>**3.30** | 2.27<br>**3.17** | 2.21<br>**3.06** | 2.16<br>**2.98** | 2.12<br>**2.90** | 2.09<br>**2.84** | 2.04<br>**2.74** | 1.99<br>**2.66** | 1.93<br>**2.55** | 1.89<br>**2.47** | 1.84<br>**2.38** | 1.79<br>**2.29** | 1.76<br>**2.24** | 1.72<br>**2.16** | 1.69<br>**2.13** | 1.66<br>**2.07** | 1.64<br>**2.03** | 1.62<br>**2.01** | 30 |
| 32 | 4.15<br>**7.50** | 3.30<br>**5.34** | 2.90<br>**4.46** | 2.67<br>**3.97** | 2.51<br>**3.66** | 2.40<br>**3.42** | 2.32<br>**3.25** | 2.25<br>**3.12** | 2.19<br>**3.01** | 2.14<br>**2.94** | 2.10<br>**2.86** | 2.07<br>**2.80** | 2.02<br>**2.70** | 1.97<br>**2.62** | 1.91<br>**2.51** | 1.86<br>**2.42** | 1.82<br>**2.34** | 1.76<br>**2.25** | 1.74<br>**2.20** | 1.69<br>**2.12** | 1.67<br>**2.08** | 1.64<br>**2.02** | 1.61<br>**1.98** | 1.59<br>**1.96** | 32 |
| 34 | 4.13<br>**7.44** | 3.28<br>**5.29** | 2.88<br>**4.42** | 2.65<br>**3.93** | 2.49<br>**3.61** | 2.38<br>**3.38** | 2.30<br>**3.21** | 2.23<br>**3.08** | 2.17<br>**2.97** | 2.12<br>**2.89** | 2.08<br>**2.82** | 2.05<br>**2.76** | 2.00<br>**2.66** | 1.95<br>**2.58** | 1.89<br>**2.47** | 1.84<br>**2.38** | 1.80<br>**2.30** | 1.74<br>**2.21** | 1.71<br>**2.15** | 1.67<br>**2.08** | 1.64<br>**2.04** | 1.61<br>**1.98** | 1.59<br>**1.94** | 1.57<br>**1.91** | 34 |
| 36 | 4.11<br>**7.39** | 3.26<br>**5.25** | 2.86<br>**4.38** | 2.63<br>**3.89** | 2.48<br>**3.58** | 2.36<br>**3.35** | 2.28<br>**3.18** | 2.21<br>**3.04** | 2.15<br>**2.94** | 2.10<br>**2.86** | 2.06<br>**2.78** | 2.03<br>**2.72** | 1.98<br>**2.62** | 1.93<br>**2.54** | 1.87<br>**2.43** | 1.82<br>**2.35** | 1.78<br>**2.26** | 1.72<br>**2.17** | 1.69<br>**2.12** | 1.65<br>**2.04** | 1.62<br>**2.00** | 1.59<br>**1.94** | 1.56<br>**1.90** | 1.55<br>**1.87** | 36 |
| 38 | 4.10<br>**7.35** | 3.25<br>**5.21** | 2.85<br>**4.34** | 2.62<br>**3.86** | 2.46<br>**3.54** | 2.35<br>**3.32** | 2.26<br>**3.15** | 2.19<br>**3.02** | 2.14<br>**2.91** | 2.09<br>**2.82** | 2.05<br>**2.75** | 2.02<br>**2.69** | 1.96<br>**2.59** | 1.92<br>**2.51** | 1.85<br>**2.40** | 1.80<br>**2.32** | 1.76<br>**2.22** | 1.71<br>**2.14** | 1.67<br>**2.08** | 1.63<br>**2.00** | 1.60<br>**1.97** | 1.57<br>**1.90** | 1.54<br>**1.86** | 1.53<br>**1.84** | 38 |
| 40 | 4.08<br>**7.31** | 3.23<br>**5.18** | 2.84<br>**4.31** | 2.61<br>**3.83** | 2.45<br>**3.51** | 2.34<br>**3.29** | 2.25<br>**3.12** | 2.18<br>**2.99** | 2.12<br>**2.88** | 2.07<br>**2.80** | 2.04<br>**2.73** | 2.00<br>**2.66** | 1.95<br>**2.56** | 1.90<br>**2.49** | 1.84<br>**2.37** | 1.79<br>**2.29** | 1.74<br>**2.20** | 1.69<br>**2.11** | 1.66<br>**2.05** | 1.61<br>**1.97** | 1.59<br>**1.94** | 1.55<br>**1.88** | 1.53<br>**1.84** | 1.51<br>**1.81** | 40 |
| 42 | 4.07<br>**7.27** | 3.22<br>**5.15** | 2.83<br>**4.29** | 2.59<br>**3.80** | 2.44<br>**3.49** | 2.32<br>**3.26** | 2.24<br>**3.10** | 2.17<br>**2.96** | 2.11<br>**2.86** | 2.06<br>**2.77** | 2.02<br>**2.70** | 1.99<br>**2.64** | 1.94<br>**2.54** | 1.89<br>**2.46** | 1.82<br>**2.35** | 1.78<br>**2.26** | 1.73<br>**2.17** | 1.68<br>**2.08** | 1.64<br>**2.02** | 1.60<br>**1.94** | 1.57<br>**1.91** | 1.54<br>**1.85** | 1.51<br>**1.80** | 1.49<br>**1.78** | 42 |
| 44 | 4.06<br>**7.24** | 3.21<br>**5.12** | 2.82<br>**4.26** | 2.58<br>**3.78** | 2.43<br>**3.46** | 2.31<br>**3.24** | 2.23<br>**3.07** | 2.16<br>**2.94** | 2.10<br>**2.84** | 2.05<br>**2.75** | 2.01<br>**2.68** | 1.98<br>**2.62** | 1.92<br>**2.52** | 1.88<br>**2.44** | 1.81<br>**2.32** | 1.76<br>**2.24** | 1.72<br>**2.15** | 1.66<br>**2.06** | 1.63<br>**2.00** | 1.58<br>**1.92** | 1.56<br>**1.88** | 1.52<br>**1.82** | 1.50<br>**1.78** | 1.48<br>**1.75** | 44 |
| 46 | 4.05<br>**7.21** | 3.20<br>**5.10** | 2.81<br>**4.24** | 2.57<br>**3.76** | 2.42<br>**3.44** | 2.30<br>**3.22** | 2.22<br>**3.05** | 2.14<br>**2.92** | 2.09<br>**2.82** | 2.04<br>**2.73** | 2.00<br>**2.66** | 1.97<br>**2.60** | 1.91<br>**2.50** | 1.87<br>**2.42** | 1.80<br>**2.30** | 1.75<br>**2.22** | 1.71<br>**2.13** | 1.65<br>**2.04** | 1.62<br>**1.98** | 1.57<br>**1.90** | 1.54<br>**1.86** | 1.51<br>**1.80** | 1.48<br>**1.76** | 1.46<br>**1.72** | 46 |
| 48 | 4.04<br>**7.19** | 3.19<br>**5.08** | 2.80<br>**4.22** | 2.56<br>**3.74** | 2.41<br>**3.42** | 2.30<br>**3.20** | 2.21<br>**3.04** | 2.14<br>**2.90** | 2.08<br>**2.80** | 2.03<br>**2.71** | 1.99<br>**2.64** | 1.96<br>**2.58** | 1.90<br>**2.48** | 1.86<br>**2.40** | 1.79<br>**2.28** | 1.74<br>**2.20** | 1.70<br>**2.11** | 1.64<br>**2.02** | 1.61<br>**1.96** | 1.56<br>**1.88** | 1.53<br>**1.84** | 1.50<br>**1.78** | 1.47<br>**1.73** | 1.45<br>**1.70** | 48 |

| Degrees of freedom for denominator | Degrees of freedom for numerator | | | | | | | | | | | | | | | | | | | | | | | | Degrees of freedom for denominator |
|---|---|---|---|---|---|---|---|---|---|---|---|---|---|---|---|---|---|---|---|---|---|---|---|---|---|
| | 1 | 2 | 3 | 4 | 5 | 6 | 7 | 8 | 9 | 10 | 11 | 12 | 14 | 16 | 20 | 24 | 30 | 40 | 50 | 75 | 100 | 200 | 500 | ∞ | |
| 50 | 4.03<br>**7.17** | 3.18<br>**5.06** | 2.79<br>**4.20** | 2.56<br>**3.72** | 2.40<br>**3.41** | 2.29<br>**3.18** | 2.20<br>**3.02** | 2.13<br>**2.88** | 2.07<br>**2.78** | 2.02<br>**2.70** | 1.98<br>**2.62** | 1.95<br>**2.56** | 1.90<br>**2.46** | 1.85<br>**2.39** | 1.78<br>**2.26** | 1.74<br>**2.18** | 1.69<br>**2.10** | 1.63<br>**2.00** | 1.60<br>**1.94** | 1.55<br>**1.86** | 1.52<br>**1.82** | 1.48<br>**1.76** | 1.46<br>**1.71** | 1.44<br>**1.68** | 50 |
| 55 | 4.02<br>**7.12** | 3.17<br>**5.01** | 2.78<br>**4.16** | 2.54<br>**3.68** | 2.38<br>**3.37** | 2.27<br>**3.15** | 2.18<br>**2.98** | 2.11<br>**2.85** | 2.05<br>**2.75** | 2.00<br>**2.66** | 1.97<br>**2.59** | 1.93<br>**2.53** | 1.88<br>**2.43** | 1.83<br>**2.35** | 1.76<br>**2.23** | 1.72<br>**2.15** | 1.67<br>**2.06** | 1.61<br>**1.96** | 1.58<br>**1.90** | 1.52<br>**1.82** | 1.50<br>**1.78** | 1.46<br>**1.71** | 1.43<br>**1.66** | 1.41<br>**1.64** | 55 |
| 60 | 4.00<br>**7.08** | 3.15<br>**4.98** | 2.76<br>**4.13** | 2.52<br>**3.65** | 2.37<br>**3.34** | 2.25<br>**3.12** | 2.17<br>**2.95** | 2.10<br>**2.82** | 2.04<br>**2.72** | 1.99<br>**2.63** | 1.95<br>**2.56** | 1.92<br>**2.50** | 1.86<br>**2.40** | 1.81<br>**2.32** | 1.75<br>**2.20** | 1.70<br>**2.12** | 1.65<br>**2.03** | 1.59<br>**1.93** | 1.56<br>**1.87** | 1.50<br>**1.79** | 1.48<br>**1.74** | 1.44<br>**1.68** | 1.41<br>**1.63** | 1.39<br>**1.60** | 60 |
| 65 | 3.99<br>**7.04** | 3.14<br>**4.95** | 2.75<br>**4.10** | 2.51<br>**3.62** | 2.36<br>**3.31** | 2.24<br>**3.09** | 2.15<br>**2.93** | 2.08<br>**2.79** | 2.02<br>**2.70** | 1.98<br>**2.61** | 1.94<br>**2.54** | 1.90<br>**2.47** | 1.85<br>**2.37** | 1.80<br>**2.30** | 1.73<br>**2.18** | 1.68<br>**2.09** | 1.63<br>**2.00** | 1.57<br>**1.90** | 1.54<br>**1.84** | 1.49<br>**1.76** | 1.46<br>**1.71** | 1.42<br>**1.64** | 1.39<br>**1.60** | 1.37<br>**1.56** | 65 |
| 70 | 3.98<br>**7.01** | 3.13<br>**4.92** | 2.74<br>**4.08** | 2.50<br>**3.60** | 2.35<br>**3.29** | 2.23<br>**3.07** | 2.14<br>**2.91** | 2.07<br>**2.77** | 2.01<br>**2.67** | 1.97<br>**2.59** | 1.93<br>**2.51** | 1.89<br>**2.45** | 1.84<br>**2.35** | 1.79<br>**2.28** | 1.72<br>**2.15** | 1.67<br>**2.07** | 1.62<br>**1.98** | 1.56<br>**1.88** | 1.53<br>**1.82** | 1.47<br>**1.74** | 1.45<br>**1.69** | 1.40<br>**1.62** | 1.37<br>**1.56** | 1.35<br>**1.53** | 70 |
| 80 | 3.96<br>**6.96** | 3.11<br>**4.88** | 2.72<br>**4.04** | 2.48<br>**3.56** | 2.33<br>**3.25** | 2.21<br>**3.04** | 2.12<br>**2.87** | 2.05<br>**2.74** | 1.99<br>**2.64** | 1.95<br>**2.55** | 1.91<br>**2.48** | 1.88<br>**2.41** | 1.82<br>**2.32** | 1.77<br>**2.24** | 1.70<br>**2.11** | 1.65<br>**2.03** | 1.60<br>**1.94** | 1.54<br>**1.84** | 1.51<br>**1.78** | 1.45<br>**1.70** | 1.42<br>**1.65** | 1.38<br>**1.57** | 1.35<br>**1.52** | 1.32<br>**1.49** | 80 |
| 100 | 3.94<br>**6.90** | 3.09<br>**4.82** | 2.70<br>**3.98** | 2.46<br>**3.51** | 2.30<br>**3.20** | 2.19<br>**2.99** | 2.10<br>**2.82** | 2.03<br>**2.69** | 1.97<br>**2.59** | 1.92<br>**2.51** | 1.88<br>**2.43** | 1.85<br>**2.36** | 1.79<br>**2.26** | 1.75<br>**2.19** | 1.68<br>**2.06** | 1.63<br>**1.98** | 1.57<br>**1.89** | 1.51<br>**1.79** | 1.48<br>**1.73** | 1.42<br>**1.64** | 1.39<br>**1.59** | 1.34<br>**1.51** | 1.30<br>**1.46** | 1.28<br>**1.43** | 100 |
| 125 | 3.92<br>**6.84** | 3.07<br>**4.78** | 2.68<br>**3.94** | 2.44<br>**3.47** | 2.29<br>**3.17** | 2.17<br>**2.95** | 2.08<br>**2.79** | 2.01<br>**2.65** | 1.95<br>**2.56** | 1.90<br>**2.47** | 1.86<br>**2.40** | 1.83<br>**2.33** | 1.77<br>**2.23** | 1.72<br>**2.15** | 1.65<br>**2.03** | 1.60<br>**1.94** | 1.55<br>**1.85** | 1.49<br>**1.75** | 1.45<br>**1.68** | 1.39<br>**1.59** | 1.36<br>**1.54** | 1.31<br>**1.46** | 1.27<br>**1.40** | 1.25<br>**1.37** | 125 |
| 150 | 3.91<br>**6.81** | 3.06<br>**4.75** | 2.67<br>**3.91** | 2.43<br>**3.44** | 2.27<br>**3.14** | 2.16<br>**2.92** | 2.07<br>**2.76** | 2.00<br>**2.62** | 1.94<br>**2.53** | 1.89<br>**2.44** | 1.85<br>**2.37** | 1.82<br>**2.30** | 1.76<br>**2.20** | 1.71<br>**2.12** | 1.64<br>**2.00** | 1.59<br>**1.91** | 1.54<br>**1.83** | 1.47<br>**1.72** | 1.44<br>**1.66** | 1.37<br>**1.56** | 1.34<br>**1.51** | 1.29<br>**1.43** | 1.25<br>**1.37** | 1.22<br>**1.33** | 150 |
| 200 | 3.89<br>**6.76** | 3.04<br>**4.71** | 2.65<br>**3.88** | 2.41<br>**3.41** | 2.26<br>**3.11** | 2.14<br>**2.90** | 2.05<br>**2.73** | 1.98<br>**2.60** | 1.92<br>**2.50** | 1.87<br>**2.41** | 1.83<br>**2.34** | 1.80<br>**2.28** | 1.74<br>**2.17** | 1.69<br>**2.09** | 1.62<br>**1.97** | 1.57<br>**1.88** | 1.52<br>**1.79** | 1.45<br>**1.69** | 1.42<br>**1.62** | 1.35<br>**1.53** | 1.32<br>**1.48** | 1.26<br>**1.39** | 1.22<br>**1.33** | 1.19<br>**1.28** | 200 |
| 400 | 3.86<br>**6.70** | 3.02<br>**4.66** | 2.62<br>**3.83** | 2.39<br>**3.36** | 2.23<br>**3.06** | 2.12<br>**2.85** | 2.03<br>**2.69** | 1.96<br>**2.55** | 1.90<br>**2.46** | 1.85<br>**2.37** | 1.81<br>**2.29** | 1.78<br>**2.23** | 1.72<br>**2.12** | 1.67<br>**2.04** | 1.60<br>**1.92** | 1.54<br>**1.84** | 1.49<br>**1.74** | 1.42<br>**1.64** | 1.38<br>**1.57** | 1.32<br>**1.47** | 1.28<br>**1.42** | 1.22<br>**1.32** | 1.16<br>**1.24** | 1.13<br>**1.19** | 400 |
| 1000 | 3.85<br>**6.66** | 3.00<br>**4.62** | 2.61<br>**3.80** | 2.38<br>**3.34** | 2.22<br>**3.04** | 2.10<br>**2.82** | 2.02<br>**2.66** | 1.95<br>**2.53** | 1.89<br>**2.43** | 1.84<br>**2.34** | 1.80<br>**2.26** | 1.76<br>**2.20** | 1.70<br>**2.09** | 1.65<br>**2.01** | 1.58<br>**1.89** | 1.53<br>**1.81** | 1.47<br>**1.71** | 1.41<br>**1.61** | 1.36<br>**1.54** | 1.30<br>**1.44** | 1.26<br>**1.38** | 1.19<br>**1.28** | 1.13<br>**1.19** | 1.08<br>**1.11** | 1000 |
| ∞ | 3.84<br>**6.64** | 2.99<br>**4.60** | 2.60<br>**3.78** | 2.37<br>**3.32** | 2.21<br>**3.02** | 2.09<br>**2.80** | 2.01<br>**2.64** | 1.94<br>**2.51** | 1.88<br>**2.41** | 1.83<br>**2.32** | 1.79<br>**2.24** | 1.75<br>**2.18** | 1.69<br>**2.07** | 1.64<br>**1.99** | 1.57<br>**1.87** | 1.52<br>**1.79** | 1.46<br>**1.69** | 1.40<br>**1.59** | 1.35<br>**1.52** | 1.28<br>**1.41** | 1.24<br>**1.36** | 1.17<br>**1.25** | 1.11<br>**1.15** | 1.00<br>**1.00** | ∞ |

## APPENDIX F
## CRITICAL VALUES OF DUNNETT'S $t$

*Instructions for use:* Locate the degrees of freedom (*df*) appropriate to your study (*df* for $MS_{within}$) in the left-hand column. Select an alpha level (.05 or .01). Then move across this row until you enter the column that corresponds to the number of conditions (including the standard or control group) in your study. If the value in the body of the table is exceeded by your obtained $t$-value, the comparison is significant at the alpha level you have selected. For example, with 4 conditions and *df* for $MS_{within} = 24$, your obtained $t$ must exceed 2.51 to be significant at .05 and must exceed 3.22 to be significant at .01.

### TWO-TAILED COMPARISONS

| | | *k = number of treatment means, including control* | | | | | | | | |
|---|---|---|---|---|---|---|---|---|---|---|
| $df_{error}$ | $\alpha$ | 2 | 3 | 4 | 5 | 6 | 7 | 8 | 9 | 10 |
| 5 | .05 | 2.57 | 3.03 | 3.29 | 3.48 | 3.62 | 3.73 | 3.82 | 3.90 | 3.97 |
| | .01 | 4.03 | 4.63 | 4.98 | 5.22 | 5.41 | 5.56 | 5.69 | 5.80 | 5.89 |
| 6 | .05 | 2.45 | 2.86 | 3.10 | 3.26 | 3.39 | 3.49 | 3.57 | 3.64 | 3.71 |
| | .01 | 3.71 | 4.21 | 4.51 | 4.71 | 4.87 | 5.00 | 5.10 | 5.20 | 5.28 |
| 7 | .05 | 2.36 | 2.75 | 2.97 | 3.12 | 3.24 | 3.33 | 3.41 | 3.47 | 3.53 |
| | .01 | 3.50 | 3.95 | 4.21 | 4.39 | 4.53 | 4.64 | 4.74 | 4.82 | 4.89 |
| 8 | .05 | 2.31 | 2.67 | 2.88 | 3.02 | 3.13 | 3.22 | 3.29 | 3.35 | 3.41 |
| | .01 | 3.36 | 3.77 | 4.00 | 4.17 | 4.29 | 4.40 | 4.48 | 4.56 | 4.62 |
| 9 | .05 | 2.26 | 2.61 | 2.81 | 2.95 | 3.05 | 3.14 | 3.20 | 3.26 | 3.32 |
| | .01 | 3.25 | 3.63 | 3.85 | 4.01 | 4.12 | 4.22 | 4.30 | 4.37 | 4.43 |
| 10 | .05 | 2.23 | 2.57 | 2.76 | 2.89 | 2.99 | 3.07 | 3.14 | 3.19 | 3.24 |
| | .01 | 3.17 | 3.53 | 3.74 | 3.88 | 3.99 | 4.08 | 4.16 | 4.22 | 4.28 |
| 11 | .05 | 2.20 | 2.53 | 2.72 | 2.84 | 2.94 | 3.02 | 3.08 | 3.14 | 3.19 |
| | .01 | 3.11 | 3.45 | 3.65 | 3.79 | 3.89 | 3.98 | 4.05 | 4.11 | 4.16 |
| 12 | .05 | 2.18 | 2.50 | 2.68 | 2.81 | 2.90 | 2.98 | 3.04 | 3.09 | 3.14 |
| | .01 | 3.05 | 3.39 | 3.58 | 3.71 | 3.81 | 3.89 | 3.96 | 4.02 | 4.07 |
| 13 | .05 | 2.16 | 2.48 | 2.65 | 2.78 | 2.87 | 2.94 | 3.00 | 3.06 | 3.10 |
| | .01 | 3.01 | 3.33 | 3.52 | 3.65 | 3.74 | 3.82 | 3.89 | 3.94 | 3.99 |
| 14 | .05 | 2.14 | 2.46 | 2.63 | 2.75 | 2.84 | 2.91 | 2.97 | 3.02 | 3.07 |
| | .01 | 2.98 | 3.29 | 3.47 | 3.59 | 3.69 | 3.76 | 3.83 | 3.88 | 3.93 |
| 15 | .05 | 2.13 | 2.44 | 2.61 | 2.73 | 2.82 | 2.89 | 2.95 | 3.00 | 3.04 |
| | .01 | 2.95 | 3.25 | 3.43 | 3.55 | 3.64 | 3.71 | 3.78 | 3.83 | 3.88 |

Table adapted from C. W. Dunnett, New tables for multiple comparisons with a control, *Biometrics*, 1964, Vol. 20, 482–491. Reprinted by permission of the author and editor.

| $df_{error}$ | α | *k = number of treatment means, including control* | | | | | | | | |
|---|---|---|---|---|---|---|---|---|---|---|
| | | 2 | 3 | 4 | 5 | 6 | 7 | 8 | 9 | 10 |
| 16 | .05 | 2.12 | 2.42 | 2.59 | 2.71 | 2.80 | 2.87 | 2.92 | 2.97 | 3.02 |
| | .01 | 2.92 | 3.22 | 3.39 | 3.51 | 3.60 | 3.67 | 3.73 | 3.78 | 3.83 |
| 17 | .05 | 2.11 | 2.41 | 2.58 | 2.69 | 2.78 | 2.85 | 2.90 | 2.95 | 3.00 |
| | .01 | 2.90 | 3.19 | 3.36 | 3.47 | 3.56 | 3.63 | 3.69 | 3.74 | 3.79 |
| 18 | .05 | 2.10 | 2.40 | 2.56 | 2.68 | 2.76 | 2.83 | 2.89 | 2.94 | 2.98 |
| | .01 | 2.88 | 3.17 | 3.33 | 3.44 | 3.53 | 3.60 | 3.66 | 3.71 | 3.75 |
| 19 | .05 | 2.09 | 2.39 | 2.55 | 2.66 | 2.75 | 2.81 | 2.87 | 2.92 | 2.96 |
| | .01 | 2.86 | 3.15 | 3.31 | 3.42 | 3.50 | 3.57 | 3.63 | 3.68 | 3.72 |
| 20 | .05 | 2.09 | 2.38 | 2.54 | 2.65 | 2.73 | 2.80 | 2.86 | 2.90 | 2.95 |
| | .01 | 2.85 | 3.13 | 3.29 | 3.40 | 3.48 | 3.55 | 3.60 | 3.65 | 3.69 |
| 24 | .05 | 2.06 | 2.35 | 2.51 | 2.61 | 2.70 | 2.76 | 2.81 | 2.86 | 2.90 |
| | .01 | 2.80 | 3.07 | 3.22 | 3.32 | 3.40 | 3.47 | 3.52 | 3.57 | 3.61 |
| 30 | .05 | 2.04 | 2.32 | 2.47 | 2.58 | 2.66 | 2.72 | 2.77 | 2.82 | 2.86 |
| | .01 | 2.75 | 3.01 | 3.15 | 3.25 | 3.33 | 3.39 | 3.44 | 3.49 | 3.52 |
| 40 | .05 | 2.02 | 2.29 | 2.44 | 2.54 | 2.62 | 2.68 | 2.73 | 2.77 | 2.81 |
| | .01 | 2.70 | 2.95 | 3.09 | 3.19 | 3.26 | 3.32 | 3.37 | 3.41 | 3.44 |
| 60 | .05 | 2.00 | 2.27 | 2.41 | 2.51 | 2.58 | 2.64 | 2.69 | 2.73 | 2.77 |
| | .01 | 2.66 | 2.90 | 3.03 | 3.12 | 3.19 | 3.25 | 3.29 | 3.33 | 3.37 |
| 120 | .05 | 1.98 | 2.24 | 2.38 | 2.47 | 2.55 | 2.60 | 2.65 | 2.69 | 2.73 |
| | .01 | 2.62 | 2.85 | 2.97 | 3.06 | 3.12 | 3.18 | 3.22 | 3.26 | 3.29 |
| ∞ | .05 | 1.96 | 2.21 | 2.35 | 2.44 | 2.51 | 2.57 | 2.61 | 2.65 | 2.69 |
| | .01 | 2.58 | 2.79 | 2.92 | 3.00 | 3.06 | 3.11 | 3.15 | 3.19 | 3.22 |

# APPENDIX G
## BINOMIAL COEFFICIENTS

*Instructions for use:* Locate the total number of observations appropriate to your study in the left-hand column (N). Move across this row until you enter the column that corresponds to the number of observations in one of the two categories ($f$). The value in the body of the table is the appropriate binomial coefficient (see text p. 184).

$$\binom{N}{f}$$

| N | $\binom{N}{0}$ | $\binom{N}{1}$ | $\binom{N}{2}$ | $\binom{N}{3}$ | $\binom{N}{4}$ | $\binom{N}{5}$ | $\binom{N}{6}$ | $\binom{N}{7}$ | $\binom{N}{8}$ | $\binom{N}{9}$ | $\binom{N}{10}$ |
|---|---|---|---|---|---|---|---|---|---|---|---|
| 0 | 1 | | | | | | | | | | |
| 1 | 1 | 1 | | | | | | | | | |
| 2 | 1 | 2 | 1 | | | | | | | | |
| 3 | 1 | 3 | 3 | 1 | | | | | | | |
| 4 | 1 | 4 | 6 | 4 | 1 | | | | | | |
| 5 | 1 | 5 | 10 | 10 | 5 | 1 | | | | | |
| 6 | 1 | 6 | 15 | 20 | 15 | 6 | 1 | | | | |
| 7 | 1 | 7 | 21 | 35 | 35 | 21 | 7 | 1 | | | |
| 8 | 1 | 8 | 28 | 56 | 70 | 56 | 28 | 8 | 1 | | |
| 9 | 1 | 9 | 36 | 84 | 126 | 126 | 84 | 36 | 9 | 1 | |
| 10 | 1 | 10 | 45 | 120 | 210 | 252 | 210 | 120 | 45 | 10 | 1 |
| 11 | 1 | 11 | 55 | 165 | 330 | 462 | 462 | 330 | 165 | 55 | 11 |
| 12 | 1 | 12 | 66 | 220 | 495 | 792 | 924 | 792 | 495 | 220 | 66 |
| 13 | 1 | 13 | 78 | 286 | 715 | 1287 | 1716 | 1716 | 1287 | 715 | 286 |
| 14 | 1 | 14 | 91 | 364 | 1001 | 2002 | 3003 | 3432 | 3003 | 2002 | 1001 |
| 15 | 1 | 15 | 105 | 455 | 1365 | 3003 | 5005 | 6435 | 6435 | 5005 | 3003 |
| 16 | 1 | 16 | 120 | 560 | 1820 | 4368 | 8008 | 11440 | 12870 | 11440 | 8008 |
| 17 | 1 | 17 | 136 | 680 | 2380 | 6188 | 12376 | 19448 | 24310 | 24310 | 19448 |
| 18 | 1 | 18 | 153 | 816 | 3060 | 8568 | 18564 | 31824 | 43758 | 48620 | 43758 |
| 19 | 1 | 19 | 171 | 969 | 3876 | 11628 | 27132 | 50388 | 75582 | 92378 | 92378 |
| 20 | 1 | 20 | 190 | 1140 | 4845 | 15504 | 38760 | 77520 | 125970 | 167960 | 184756 |

Reproduced by permission from Richard P. Runyon and Audrey Haber, *Fundamentals of Behavioral Statistics*, 3rd Ed. Reading, Mass.: Addison-Wesley, 1976, p. 397.

## APPENDIX H
## CRITICAL VALUES OF CHI-SQUARE

*Instructions for use:* Locate the degrees of freedom (*df*) appropriate to your study in the left-hand column. Then move across this row until you enter the column that corresponds to your alpha level. If the value in the body of the table is exceeded by your obtained value of Chi-square, it is significant at the alpha level you have selected. For example, with $df = 1$, your obtained value must exceed 6.64 to be significant at the .01 level.

| *df* | .20 | .10 | .05 | .02 | .01 | .001 |
|---|---|---|---|---|---|---|
| 1 | 1.64 | 2.71 | 3.84 | 5.41 | 6.64 | 10.83 |
| 2 | 3.22 | 4.60 | 5.99 | 7.82 | 9.21 | 13.82 |
| 3 | 4.64 | 6.25 | 7.82 | 9.84 | 11.34 | 16.27 |
| 4 | 5.99 | 7.78 | 9.49 | 11.67 | 13.28 | 18.46 |
| 5 | 7.29 | 9.24 | 11.07 | 13.39 | 15.09 | 20.52 |
| 6 | 8.56 | 10.64 | 12.59 | 15.03 | 16.81 | 22.46 |
| 7 | 9.80 | 12.02 | 14.07 | 16.62 | 18.48 | 24.32 |
| 8 | 11.03 | 13.36 | 15.51 | 18.17 | 20.09 | 26.12 |
| 9 | 12.24 | 14.68 | 16.92 | 19.68 | 21.67 | 27.88 |
| 10 | 13.44 | 15.99 | 18.31 | 21.16 | 23.21 | 29.59 |
| 11 | 14.63 | 17.28 | 19.68 | 22.62 | 24.72 | 31.26 |
| 12 | 15.81 | 18.55 | 21.03 | 24.05 | 26.22 | 32.91 |
| 13 | 16.98 | 19.81 | 22.36 | 25.47 | 27.69 | 34.53 |
| 14 | 18.15 | 21.06 | 23.68 | 26.87 | 29.14 | 36.12 |
| 15 | 19.31 | 22.31 | 25.00 | 28.26 | 30.58 | 37.70 |
| 16 | 20.46 | 23.54 | 26.30 | 29.63 | 32.00 | 39.25 |
| 17 | 21.62 | 24.77 | 27.59 | 31.00 | 33.41 | 40.79 |
| 18 | 22.76 | 25.99 | 28.87 | 32.35 | 34.80 | 42.31 |
| 19 | 23.90 | 27.20 | 30.14 | 33.69 | 36.19 | 43.82 |
| 20 | 25.04 | 28.41 | 31.41 | 35.02 | 37.57 | 45.32 |
| 21 | 26.17 | 29.62 | 32.67 | 36.34 | 38.93 | 46.80 |
| 22 | 27.30 | 30.81 | 33.92 | 37.66 | 40.29 | 48.27 |
| 23 | 28.43 | 32.01 | 35.17 | 38.97 | 41.64 | 49.73 |
| 24 | 29.55 | 33.20 | 36.42 | 40.27 | 42.98 | 51.18 |
| 25 | 30.68 | 34.38 | 37.65 | 41.57 | 44.31 | 52.62 |
| 26 | 31.80 | 35.56 | 38.88 | 42.86 | 45.64 | 54.05 |
| 27 | 32.91 | 36.74 | 40.11 | 44.14 | 46.96 | 55.48 |
| 28 | 34.03 | 37.92 | 41.34 | 45.42 | 48.28 | 56.89 |
| 29 | 35.14 | 39.09 | 42.56 | 46.69 | 49.59 | 58.30 |
| 30 | 36.25 | 40.26 | 43.77 | 47.96 | 50.89 | 59.70 |

Reproduced with permission from Q. McNemar, *Psychological Statistics,* 3rd Ed. New York: John Wiley and Sons, Inc., 1962.

# APPENDIX I
## CRITICAL VALUES OF $U$

*Instructions for use:* Select the table which corresponds to your alpha level (.05 or .01). Find the $N$ for the smaller of your two samples ($N_1$) in the left-hand column. Move across the row until you enter the column corresponding to $N$ in the larger of your two samples ($N_2$). (If the two samples have equal $N$, this decision is arbitrary). Your obtained $U$ is significant only if it does *not* exceed the value you have located in the body of the table.

Two-Tailed Test, $\alpha$ = .05

| | $N_2$ | | | | | | | | | | | | | | | | | |
|---|---|---|---|---|---|---|---|---|---|---|---|---|---|---|---|---|---|---|
| $N_1$ | 3 | 4 | 5 | 6 | 7 | 8 | 9 | 10 | 11 | 12 | 13 | 14 | 15 | 16 | 17 | 18 | 19 | 20 |
| 1 | – | – | – | – | – | – | – | – | – | – | – | – | – | – | – | – | – | – |
| 2 | – | – | – | – | – | 0 | 0 | 0 | 0 | 1 | 1 | 1 | 1 | 1 | 2 | 2 | 2 | 2 |
| 3 | – | – | 0 | 1 | 1 | 2 | 2 | 3 | 3 | 4 | 4 | 5 | 5 | 6 | 6 | 7 | 7 | 8 |
| 4 | | 0 | 1 | 2 | 3 | 4 | 4 | 5 | 6 | 7 | 8 | 9 | 10 | 11 | 11 | 12 | 13 | 13 |
| 5 | | | 2 | 3 | 5 | 6 | 7 | 8 | 9 | 11 | 12 | 13 | 14 | 15 | 17 | 18 | 19 | 20 |
| 6 | | | | 5 | 6 | 8 | 10 | 11 | 13 | 14 | 16 | 17 | 19 | 21 | 22 | 24 | 25 | 27 |
| 7 | | | | | 8 | 10 | 12 | 14 | 16 | 18 | 20 | 22 | 24 | 26 | 28 | 30 | 32 | 34 |
| 8 | | | | | | 13 | 15 | 17 | 19 | 22 | 24 | 26 | 29 | 31 | 34 | 36 | 38 | 41 |
| 9 | | | | | | | 17 | 20 | 23 | 26 | 28 | 31 | 34 | 37 | 39 | 42 | 45 | 48 |
| 10 | | | | | | | | 23 | 26 | 29 | 33 | 36 | 39 | 42 | 45 | 48 | 52 | 55 |
| 11 | | | | | | | | | 30 | 33 | 37 | 40 | 44 | 47 | 51 | 55 | 58 | 62 |
| 12 | | | | | | | | | | 37 | 41 | 45 | 49 | 53 | 57 | 61 | 65 | 69 |
| 13 | | | | | | | | | | | 45 | 50 | 54 | 59 | 63 | 67 | 72 | 76 |
| 14 | | | | | | | | | | | | 55 | 59 | 64 | 67 | 74 | 78 | 83 |
| 15 | | | | | | | | | | | | | 64 | 70 | 75 | 80 | 85 | 90 |
| 16 | | | | | | | | | | | | | | 75 | 81 | 86 | 92 | 98 |
| 17 | | | | | | | | | | | | | | | 87 | 93 | 99 | 105 |
| 18 | | | | | | | | | | | | | | | | 99 | 106 | 112 |
| 19 | | | | | | | | | | | | | | | | | 113 | 119 |
| 20 | | | | | | | | | | | | | | | | | | 127 |

Reprinted by permission from L. C. Freeman, *Elementary Applied Statistics.* New York: John Wiley and Sons, Inc., 1965.

Two-Tailed Test, $\alpha = .01$

| | $N_2$ | | | | | | | | | | | | | | | | | |
|---|---|---|---|---|---|---|---|---|---|---|---|---|---|---|---|---|---|---|
| $N_1$ | 3 | 4 | 5 | 6 | 7 | 8 | 9 | 10 | 11 | 12 | 13 | 14 | 15 | 16 | 17 | 18 | 19 | 20 |
| 1 | – | – | – | – | – | – | – | – | – | – | – | – | – | – | – | – | – | – |
| 2 | – | – | – | – | – | – | – | – | – | – | – | – | – | – | – | – | 0 | 0 |
| 3 | – | – | – | – | – | – | 0 | 0 | 0 | 1 | 1 | 1 | 2 | 2 | 2 | 2 | 3 | 3 |
| 4 | | – | – | 0 | 0 | 1 | 1 | 2 | 2 | 3 | 4 | 4 | 5 | 5 | 6 | 6 | 7 | 8 |
| 5 | | | 0 | 1 | 2 | 3 | 3 | 4 | 5 | 6 | 7 | 7 | 8 | 9 | 10 | 11 | 12 | 13 |
| 6 | | | | 2 | 3 | 4 | 5 | 6 | 7 | 9 | 10 | 11 | 12 | 13 | 15 | 16 | 17 | 18 |
| 7 | | | | | 4 | 6 | 7 | 9 | 10 | 12 | 13 | 15 | 16 | 18 | 19 | 21 | 22 | 24 |
| 8 | | | | | | 8 | 10 | 12 | 14 | 16 | 18 | 19 | 21 | 23 | 25 | 27 | 29 | 31 |
| 9 | | | | | | | 11 | 13 | 16 | 18 | 20 | 22 | 24 | 27 | 29 | 31 | 33 | 36 |
| 10 | | | | | | | | 16 | 18 | 21 | 24 | 26 | 29 | 31 | 34 | 37 | 39 | 42 |
| 11 | | | | | | | | | 21 | 24 | 27 | 30 | 33 | 36 | 39 | 42 | 45 | 48 |
| 12 | | | | | | | | | | 27 | 31 | 34 | 37 | 41 | 44 | 47 | 51 | 54 |
| 13 | | | | | | | | | | | 34 | 38 | 42 | 45 | 49 | 53 | 56 | 60 |
| 14 | | | | | | | | | | | | 42 | 46 | 50 | 54 | 58 | 63 | 67 |
| 15 | | | | | | | | | | | | | 51 | 55 | 60 | 64 | 69 | 73 |
| 16 | | | | | | | | | | | | | | 60 | 65 | 70 | 74 | 79 |
| 17 | | | | | | | | | | | | | | | 70 | 75 | 81 | 86 |
| 18 | | | | | | | | | | | | | | | | 81 | 87 | 92 |
| 19 | | | | | | | | | | | | | | | | | 93 | 99 |
| 20 | | | | | | | | | | | | | | | | | | 105 |

## APPENDIX J
## CRITICAL VALUES OF RHO

*Instructions for use:* Locate the number of observations (difference scores) appropriate to your study in the left-hand column ($N$). Move across this row until you enter the column that corresponds to your selected alpha level. If the value in the body of the table is exceeded by your obtained value of rho, it is significant at the alpha level you have selected.

| | *Level of Significance* $\alpha$ | | | | |
|---|---|---|---|---|---|
| $N$ | .20 | .10 | .05 | .02 | .01 |
| 4 | | 1.00 | | | |
| 5 | .80 | .90 | | 1.00 | |
| 6 | .66 | .83 | .89 | .94 | 1.00 |
| 7 | .57 | .71 | .79 | .89 | .93 |
| 8 | .52 | .64 | .74 | .83 | .88 |
| 9 | .48 | .60 | .68 | .78 | .83 |
| 10 | .45 | .56 | .65 | .73 | .79 |
| 11 | .41 | .52 | .61 | .71 | .77 |
| 12 | .39 | .50 | .59 | .68 | .75 |
| 13 | .37 | .47 | .56 | .65 | .71 |
| 14 | .36 | .46 | .54 | .63 | .69 |
| 15 | .34 | .44 | .52 | .60 | .66 |
| 16 | .33 | .42 | .51 | .58 | .64 |
| 17 | .32 | .41 | .49 | .57 | .62 |
| 18 | .31 | .40 | .48 | .55 | .61 |
| 19 | .30 | .39 | .46 | .54 | .60 |
| 20 | .29 | .38 | .45 | .53 | .58 |
| 21 | .29 | .37 | .44 | .51 | .56 |
| 22 | .28 | .36 | .43 | .50 | .55 |
| 23 | .27 | .35 | .42 | .49 | .54 |
| 24 | .27 | .34 | .41 | .48 | .53 |
| 25 | .26 | .34 | .40 | .47 | .52 |
| 26 | .26 | .33 | .39 | .46 | .51 |
| 27 | .25 | .32 | .38 | .45 | .50 |
| 28 | .25 | .32 | .38 | .44 | .49 |
| 29 | .24 | .31 | .37 | .44 | .48 |
| 30 | .24 | .31 | .36 | .43 | .47 |

Reproduced by permission from E. G. Old, Distribution of sums of squares of rank differences, and The 5% significance levels for sums of squares of rank differences and a correction, *Annals of Mathematical Statistics*, 9: 133–148, 1938 and 20: 117–118, 1949.

# Index

## C



## D



## E



## F

## G



## H



## I



## L



## M



## N



## O



## P

## U



## V



## W



## XYZ